# Geological Storage and Monitoring of Carbon Dioxide

Dong Li · Yinling Guo

# Geological Storage and Monitoring of Carbon Dioxide

Springer

Dong Li
School of Resource and Safety Engineering
University of Science and Technology Beijing
Beijing, China

Yinling Guo
State Key Lab for Fine Exploration and Intelligent Development of Coal Resources
China University of Mining and Technology (Beijing)
Beijing, China

ISBN 978-981-92-3620-6 ISBN 978-981-92-3621-3 (eBook)
https://doi.org/10.1007/978-981-92-3621-3

The original submitted manuscript has been translated into English. The translation was done using artificial intelligence. A subsequent revision was performed by the author(s) to further refine the work and to ensure that the translation is appropriate concerning content and scientific correctness. It may, however, read stylistically different from a conventional translation.



This Springer imprint is published by the registered company Springer Nature Singapore Pte Ltd.
The registered company address is: 152 Beach Road, #21-01/04 Gateway East, Singapore 189721, Singapore

# Preface

In response to global climate change, carbon neutrality has become China's national strategy for low-carbon development over the next 40 years, which will have a profound impact on the country's energy structure and economic system. Large-scale coal exploitation and utilisation is the main source of $CO_2$ emissions in China. However, given China's coal-dominated energy structure, coal will remain the "cornerstone" of energy supply for some time to come. As a result, China faces enormous pressure to reduce carbon emissions. Against the backdrop of balancing economic development with carbon emission targets, negative carbon technologies will be a crucial pathway for China to achieve carbon neutrality. Carbon Capture, Utilisation and Storage (CCUS) is one of the fastest and most effective negative carbon technologies for reducing atmospheric $CO_2$ concentrations. According to international energy forecasts, under sustainable development goals, CCUS will contribute 15% of the cumulative global carbon emission reductions required to achieve net-zero $CO_2$ emissions by 2070. Therefore, based on China's coal-based energy structure and resource endowment, geological storage of $CO_2$ can enhance the flexibility of the energy system and enable low-carbon development during the transition away from fossil energy, making it a fundamental technology for China to achieve carbon neutrality.

Geophysical exploration, as a technique for understanding subsurface geological conditions, has achieved mature application results in resource exploration and development. With the exploration and implementation of $CO_2$ geological storage projects, new application scenarios have emerged for geophysical exploration, further broadening and extending the development of geophysical exploration technologies. In recent years, this field has attracted widespread attention from both academics and engineering professionals, bringing not only new opportunities but also new challenges.

From a geological perspective, achieving long-term and effective $CO_2$ storage requires not only favourable geological properties of the storage formation, but also the assurance of stable containment without leakage following $CO_2$ injection. Favourable storage properties include: suitable formation porosity, permeability, pore pressure, an intact and stable caprock, and the absence of interconnected faults, subsidence columns, fractures, and other potential leakage pathways. Therefore, it is essential to understand the lithological, physical, and structural characteristics of the formations, as well as the migration behaviour of $CO_2$ at different stages of injection. However, $CO_2$ geological storage in China remains at the demonstration project stage, and a mature and comprehensive technological system has yet to be established. Traditional geophysical exploration methods, which are primarily designed for resource exploration, struggle to meet the high-resolution, high-precision, and high-efficiency requirements of $CO_2$ geological storage. Furthermore, there is a shift from "short-term" resource exploration to "long-term" monitoring and early warning. In response to the theoretical and technical issues related to $CO_2$ geological storage, the author has compiled this book, *Geological Storage and Monitoring of Carbon Dioxide*. This book focuses on the core aspects of $CO_2$ geological storage theory, technology, and engineering practice, providing a detailed exposition of the significance, current status, theories, and methodologies of $CO_2$ geological storage. It introduces new geophysical exploration and monitoring technologies for different stages of $CO_2$ geological storage, analyses the practical application effects of these new methods, and aims to provide a reference for the implementation of $CO_2$ geological storage projects.

This book focuses on highlighting the application of geophysical exploration techniques in $CO_2$ geological storage, discussing the contributions of geophysical methods from the perspectives of safety and effectiveness assessment. It aims to provide new insights for students specialising in geophysics and related researchers, breaking through traditional disciplinary boundaries and emphasising interdisciplinary collaboration. We also hope that readers will find inspiration and support from this book, and that together we can contribute to achieving the goal of carbon neutrality.

This book has received substantial support in terms of practical data from the National Key Laboratory for Precision Exploration and Intelligent Development of Coal Resources. During the writing process, we also benefited from multiple reviews and suggestions provided by Wang Guanjun, Chen Longquan, Wang Shiqi, and others. A large number of books and literature sources were consulted in the preparation of this book, some of which have been listed in the references. We would like to express our sincere gratitude to the authors of these works, as well as to those whose contributions have not been explicitly cited.

Due to the limitations of the author's expertise, shortcomings in this book are inevitable. We warmly welcome feedback and corrections from readers and fellow experts, so that we may continually rectify errors, draw on valuable experience, and further improve and enhance this work.

Beijing, China<br>December 2025

Dong Li<br>Yinling Guo

# Summary

This book focuses on geological storage monitoring of $CO_2$ and related geological tasks. It systematically presents the significance, mechanisms, technologies, and methods of $CO_2$ geological storage, aiming to enhance readers' understanding of the safety and effectiveness of $CO_2$ geological storage. The specific content includes an overview of $CO_2$ geological storage, theoretical and mechanistic analysis, numerical simulation techniques, the contribution of geophysical technologies to $CO_2$ geological storage, characterisation techniques for $CO_2$ storage reservoirs, monitoring technologies for $CO_2$ migration, and well-ground integrated microseismic monitoring technologies for the safety of $CO_2$ geological storage. The book progresses from basic to advanced topics, introducing relevant content step by step from the "significance–pathways–theory–methods" of $CO_2$ geological storage. It emphasizes the role of geophysical technologies in the safety and effectiveness assessment of $CO_2$ geological storage, enabling readers to grasp fundamental theories, understand current developments, and assimilate relevant technologies.

This book can be used as a textbook for students majoring in geophysics, geology, earth exploration and information technology, carbon neutrality, and related disciplines. It may also serve as a reference for researchers and engineering professionals in relevant fields.

# Contents

# Chapter 1
# Overview of Geological Storage of Carbon Dioxide

Since China acceded to the Paris Agreement, it has become an important participant and contributor in the global response to climate change. The 28th Conference of the Parties (COP28) to the United Nations Framework Convention on Climate Change, held in 2023, conducted a global stocktake of the implementation of the Paris Agreement. As a developing country, China, while balancing economic development with carbon emission targets, has implemented a series of measures and actions to address climate change, achieving positive results in this regard.

Geological storage of $CO_2$ refers to the process of capturing and separating $CO_2$ from stationary carbon emission sources, compressing and liquefying it, and subsequently using it for enhanced oil recovery, enhanced gas recovery, or directly storing it in underground geological formations or structures, thereby achieving long-term isolation of $CO_2$ from the atmosphere. China is still at a critical stage of economic development, and there remains considerable short-term demand for fossil energy. Therefore, in balancing economic development with carbon emission targets, geological storage of $CO_2$ is a fundamental technology for China to achieve its carbon neutrality goals.

## 1.1 Global Climate Change and Response Measures

Global climate is undergoing unprecedented changes, with climate warming caused by greenhouse gas emissions attracting widespread attention. International experts have conducted in-depth analyses and discussions on global climate change, ranging from the causes of these changes to trend predictions. Preliminary conclusions and response measures have been formulated, which have also elicited active responses from various countries.



D. Li and Y. Guo, *Geological Storage and Monitoring of Carbon Dioxide*,
https://doi.org/10.1007/978-981-92-3621-3_1

### *1.1.1 Global Climate Change*

Greenhouse gases play a crucial role in maintaining the Earth's thermal equilibrium. Under the influence of greenhouse gases such as water vapour and $CO_2$ in the atmosphere, a temperature environment most suitable for life on Earth is established, thereby enabling the survival and flourishing of living organisms. In the absence of the atmospheric layer and greenhouse gases such as water vapour and $CO_2$ within it, the Earth's surface temperature would be 33 °C lower than at present, posing a survival crisis for humans and the vast majority of flora and fauna. However, excessive emissions of greenhouse gases resulting from human activities have led to an increased retention of terrestrial thermal radiation, disrupting the thermal balance and consequently triggering global climate warming.

On 9 August 2021, the Intergovernmental Panel on Climate Change (IPCC) released the report of Working Group I of its Sixth Assessment Report (AR6). The IPCC AR6 points out that over the past century, and especially since the 1970s, the global climate has been experiencing significant and rapid warming; from 2011 to 2020, the global surface temperature was approximately 1.09 °C higher than the pre-industrial level (the 1850–1900 average). The main driver of global warming is the emission of greenhouse gases such as $CO_2$ from human activities. Among these, $CO_2$ accounts for about 66% of the radiative forcing from long-lived greenhouse gases and contributes approximately 82% to the increase in radiative forcing over the past decade. In 2019, the atmospheric concentration of $CO_2$ reached 410.5 mg/L, which is 148% of the pre-industrial level (1750), representing the highest value at any time in the past two million years. Therefore, controlling the concentration of greenhouse gases in the atmosphere has become a central issue in successive rounds of international climate negotiations. The IPCC assessment report also notes that the observed average temperature increase over land surfaces is nearly twice that of the global mean surface temperature (GMST), which includes both land and ocean. It is noteworthy that since 1971, the oceans, which cover about 71% of the Earth's surface, have absorbed approximately 93% of the excess heat generated by greenhouse gas emissions from human activities. Between 1993 and 2017, both the rate of ocean heat uptake and the total heat content were roughly double the average levels observed between 1970 and 1993.

Observational evidence indicates that the impacts of climate warming are already manifesting across continents, oceans, and human societies. In many regions, changes in precipitation patterns or the melting of ice and snow are altering hydrological systems, thereby affecting both the quantity and quality of water resources. Numerous terrestrial and freshwater species have shifted their geographical distributions, seasonal activities, and migration patterns. Furthermore, global warming has led to changes in climate zones, land degradation, and desertification in many areas, posing risks to food security, livelihoods, infrastructure, land values, as well as to ecosystems and human health. Similarly, the oceans are experiencing severe impacts and risks associated with climate change, including reductions in marine biodiversity, shifts in species distributions, and declines in productivity. For example, sea level rise, ocean warming, acidification, deoxygenation, and changes in nutrient availability are reshaping the structure and function of marine ecosystems, posing significant threats to the safety and sustainable development of human societies. Despite the increasingly evident and severe consequences of climate warming and its associated risks, the international community appears to be inadequately prepared to respond effectively. Therefore, strengthening responses to the impacts and risks of climate change—including controlling greenhouse gas emissions, enhancing adaptive capacity, and safeguarding the sustainable development of economies and societies—has become a major challenge currently facing the international community.

Since the 1950s, the concentration of $CO_2$ in the atmosphere has increased year by year (Fig. 1.1), with anthropogenic activities causing carbon emissions to far exceed the Earth system's natural carbon sink capacity. Excessive emissions of greenhouse gases, represented by $CO_2$, have led to a continuous rise in global average temperatures (Fig. 1.2). The resulting global climate change has triggered a series of extreme high-temperature events, such as glacier melt, sea level rise, ocean acidification, and ecosystem collapse, which are having a profound impact on the natural ecological environment and pose significant threats to human economic and social development. In 2024, the global average temperature is approximately 1.50 °C higher than the pre-industrial average (1850–1900). Between 1951 and 2020, the average annual surface temperature in China increased by 0.26 °C per decade, significantly higher than the global average increase of 0.15 °C per decade over the same period. In 2021, the United Nations Climate Change Conference set the goal of "limiting global warming to 1.5 °C by the end of this century" as one of the key objectives to ensure the long-term survival of humanity on Earth, and has made comprehensive efforts to promote the transition of energy systems towards low-carbon and carbon-free development. Reducing emissions of the greenhouse gas $CO_2$ to address climate change has gradually become an international consensus, and China has also identified this as an important approach to achieving the clean and efficient utilisation of fossil energy sources such as coal.

### *1.1.2 Measures to Address Atmospheric Variation*

In order to address the impacts and threats posed by climate change, the international community, drawing on scientific foundations such as the IPCC AR5, adopted the Paris Agreement in December 2015, thereby establishing a framework for tackling climate change. The Agreement set out the global targets for controlling climate warming in this century, namely to limit the rise in global average temperature to well below 2 °C above pre-industrial levels (1850–1900 average), and to pursue efforts to limit the temperature increase to 1.5 °C. Consequently, how to control global anthropogenic emissions of greenhouse gases such as $CO_2$ has become a pressing and significant issue for the international community.

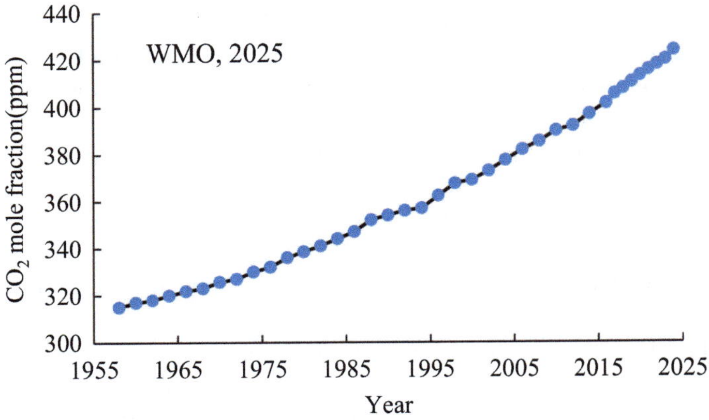


**Fig. 1.1** Variation in atmospheric $CO_2$ concentration

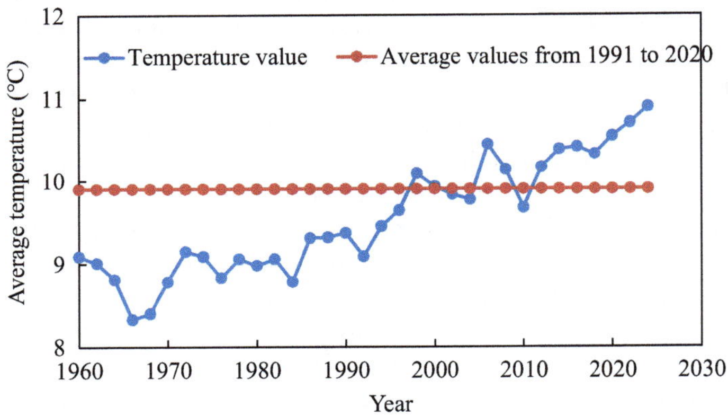


**Fig. 1.2** Interannual variation in average temperature

According to the decision of the Paris Agreement, the United Nations Framework Convention on Climate Change (UNFCCC) Secretariat invited the IPCC to prepare the Special Report on Global Warming of 1.5 °C (SR15). This special report, published in October 2018, compares the impacts and risks associated with global warming of 1.5 and 2 °C, and proposes pathways for controlling greenhouse gas emissions. In order to achieve the global temperature rise limit of 2 °C (1.5 °C) this century, global $CO_2$ emission reduction pathways require that net anthropogenic $CO_2$ emissions be reduced by approximately 25% (45%) from 2010 levels by 2030, and reach net zero emissions around 2070 (2050). In other words, anthropogenic sources and sinks of $CO_2$ and other greenhouse gases must be balanced, achieving carbon neutrality. Therefore, global emissions of $CO_2$ and other greenhouse gases need to peak as soon as possible between 2020 and 2030, i.e., reach carbon peaking.

In August 2021, the IPCC released its Sixth Assessment Report (AR6), which further indicated that under a medium greenhouse gas emissions scenario (SSP2-4.5), by the end of this century, global warming is extremely likely (95–100%) to exceed 2 °C relative to pre-industrial levels, and likely (> 66–100%) to exceed 1.5 °C in the near term (2021–2040). Only under low or very low emissions scenarios (SSP1-2.6 or SSP1-1.9) is it unlikely (0–33%) or extremely unlikely (0–5%) that global warming will exceed 2 °C by the end of the century; however, it remains more likely than not (> 50–100%) to exceed 1.5 °C. Therefore, according to the latest assessment results from the IPCC SR15 and AR6, the international community must undertake further and deeper emission reductions in the coming decades in order to limit global warming to 2 °C, or even 1.5 °C, by the end of this century.

Nevertheless, fossil fuels remain the primary energy source for human society. Therefore, achieving carbon neutrality while ensuring stable economic and social development is a formidable challenge facing the international community. By 2019, many developed countries and regions, as well as a few developing countries or regions that had experienced rapid economic growth, had already reached their carbon emission peaks and are relatively well positioned to achieve carbon neutrality in the future. However, as the largest developing country, China's carbon emissions are still on the rise, and the challenges it faces are particularly pronounced. Despite this, in September 2020, China announced its commitment to strive for $CO_2$ emissions to peak before 2030 and to achieve carbon neutrality before 2060. This is not only China's proposal for participating in global climate governance, but also a concrete manifestation of China's commitment to the concept of a shared future for humanity.

## 1.2 Carbon Dioxide Geological Storage Methods

Geological storage methods for $CO_2$ include $CO_2$ enhanced coal bed methane recovery ($CO_2$ ECBM), $CO_2$ enhanced oil recovery ($CO_2$ EOR), and $CO_2$ saline aquifer storage, among others. Overall, among the main types of geological storage, saline aquifers offer the greatest storage potential and can accommodate the largest scale of carbon storage. Oil and gas reservoir storage has the most extensive engineering practice, the best economic viability, and has already achieved initial commercialisation, with project scales in China reaching up to 1 million tonnes per year. For deep coal seam storage, China has also implemented several engineering demonstration projects; however, the current injection scale is much smaller than that of oil and gas reservoir storage, with individual project $CO_2$ injection volumes not exceeding 5000 tonnes. Nevertheless, coal seam storage offers the highest level of safety, favourable source-sink matching conditions, and good economic prospects, indicating significant potential for large-scale geological storage in the future.

## *1.2.1 Sequestration Mechanism*

The main $CO_2$ geological storage mechanisms include four types: structural trapping, residual trapping, solubility trapping, and mineral trapping, as shown in Fig. 1.3.

Structural trapping refers to the process whereby, after $CO_2$ enters the geological formation, it is retained within the pore spaces under the action of capillary forces. Residual trapping occurs very rapidly; a porous reservoir rock can be likened to a sponge—when gas is injected into the formation, it displaces the fluids within the porous rock. As the $CO_2$ continues to migrate, further displacement takes place. However, some $CO_2$ will, like water in a sponge, remain trapped within the pore spaces.

Free-phase trapping refers to the process whereby $CO_2$, after being injected into a geological formation for storage, is retained in the pore spaces of the geological medium in a gaseous state, thereby achieving storage. However, under the temperature and pressure conditions of the subsurface, $CO_2$ predominantly exists in a supercritical state rather than as an ideal gas phase.

Dissolution trapping refers to the process whereby, after $CO_2$ is injected into a geological formation, it dissolves upon contact with formation brine. The majority of the dissolved $CO_2$ exists as relatively stable carbonate ions. Following dissolution, the density of the brine in the upper layer increases. When the thickness of this high-density brine layer reaches a certain threshold, the internal pressure equilibrium of the brine is disrupted, resulting in the phenomenon known as "fingering." Under the

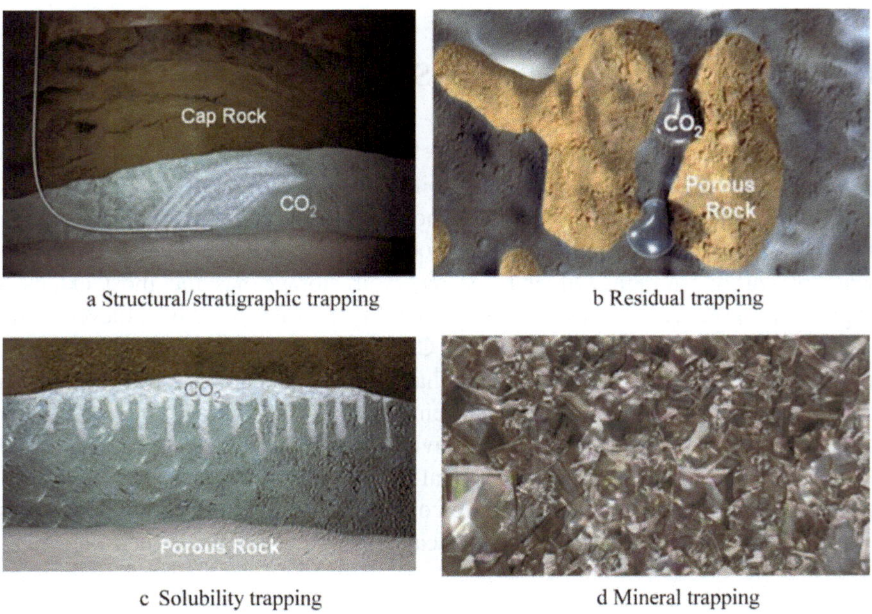


**Fig. 1.3** Mechanisms of $CO_2$ geological storage

influence of gravity, the high-density brine migrates and settles downward, reducing the presence of gaseous $CO_2$ within the formation. Meanwhile, the low-density brine flows upward, creating internal circulation within the sealed system and gradually dissolving more $CO_2$. Compared with the previous two trapping mechanisms, $CO_2$ in the dissolved state offers a higher degree of security.

Mineral storage involves the chemical reaction between $CO_2$ dissolved in water and minerals such as olivine and serpentine present in geological formations, resulting in the formation of certain clay minerals. These clay minerals remain stably sequestered in the strata over the long term. Therefore, mineral storage offers the highest level of safety.

### 1.2.2 Calculation of Storage Capacity

To establish a calculation model for the effective storage capacity of $CO_2$, it is generally necessary to consider the following aspects: the dissolution of $CO_2$ in formation water, the dissolution of $CO_2$ in crude oil, the storage of $CO_2$ in effective reservoir pore space, and the chemical trapping by reservoir minerals.

(1) Dissolution of $CO_2$ in Crude Oil

The formula for the storage capacity of $CO_2$ in oil reservoirs is as follows:

$$M_t = \rho_r[C_{os}(1 - E_R)Ah\phi(1 - S_{wi})] \tag{1.1}$$

In the formula,

- $M_t$ theoretical $CO_2$ storage capacity in the oil reservoir, $10^6$ t;
- $\rho_r$ density of $CO_2$ under reservoir conditions, kg/m$^3$;
- $C_{os}$ $CO_2$ solubility coefficient in oil;
- $A$ oil storage area, m$^2$;
- $h$ oil reservoir thickness, m;
- $\phi$ oil-impregnated porosity;
- $S_{wi}$ oil-bearing core water saturation;
- $E_R$ the crude oil recovery rate.

The solubility in oil can be calculated using the following method:

Assuming that gas dissolution is a surface chemical reaction, the main components of crude oil—namely saturates, resins, and asphaltenes—can be considered to have a negligible effect on gas dissolution. Under certain conditions, however, saturates, resins, and asphaltenes may all potentially dissolve in crude oil.

The surface chemical reaction equation is as follows:

$$G(g) + O_T(l) \rightleftharpoons GO_T(l) + \Delta H \tag{1.2}$$

The above reaction reaches dissolution equilibrium under certain temperature and pressure conditions, and the equilibrium constant of the reaction is:

$$K = \frac{[GO_T]}{p[O_T]} = \frac{\frac{n_g}{V_T}}{p\frac{n_T - n_g}{V_T}} = \frac{n_g}{p(n_T - n_g)} \tag{1.3}$$

In the formula,

$[GO_T]$ equilibrium concentration of the composite, mol;
$[O_T]$ equilibrium concentration of the gas, mol;
$p$ compressive strength, MPa;
$n_T$ the amount of substance of the liquid fuel, mol;
$n_g$ the amount of substance of the fluid gas, mol;
$V_T$ volume of liquid fuel burnt, cm$^3$;
$K$ equilibrium constant of the reaction.

In order to simplify the expression for solubility and avoid higher-order terms, a correction factor $\alpha$ is introduced into the activity coefficient term, namely:

$$[O_T] = \frac{n_T - \alpha n_g}{V_T} \tag{1.4}$$

Substituting Eq. (1.4) into Eq. (1.3), we obtain:

$$K = \frac{n_g}{p(n_T - \alpha n_g)} \tag{1.5}$$

Based on the above fundamental assumptions, Eq. (1.6) is obtained:

$$n_g = \frac{Kpn_T}{1 + \alpha Kp} \tag{1.6}$$

By subtracting $V_2$ from both sides, we obtain:

$$C = \frac{n_g}{V_2} = \frac{Kpn_T}{1 + \alpha KpV_2} \tag{1.7}$$

where

$C$ molar solubility, mol/cm$^3$;
$V_2$ the volume, in cm$^3$, of crude oil at pressure $p$ when dissolution equilibrium is reached.

The volume of combustion is:

$$V_T = \frac{n_T M_T}{\rho_T} = \frac{\rho_o V_o x_T}{\rho_T} \tag{1.8}$$

By rearranging Eq. (1.8), we obtain:

$$n_T = \frac{V_T \rho_T}{M_T} \tag{1.9}$$

In the formula,

$\rho_T$ the density of wax in crude oil, g/cm$^3$;
$M_T$ the average relative molecular mass of the gas;
$V_T$ volume of the crucible, cm$^3$;
$V_o$ the volume of crude oil under surface conditions, cm$^3$;
$\rho_o$ crude oil density, g/cm$^3$;
$x_T$ mass fraction of ash, decimal.

According to Eqs. (1.8) and (1.9), Eq. (1.7) can be rearranged to yield the theoretical model for calculating the solubility of a single-component gas:

$$\begin{aligned} C &= \frac{Kp}{1+\alpha Kp} \cdot \frac{n_T}{V_2} = \frac{Kp}{1+\alpha Kp} \cdot \frac{\rho_T V_T}{M_T V_2} = \frac{Kp}{1+\alpha Kp} \cdot \frac{\rho_o x_T}{M_T} \frac{1}{B} \\ &= \frac{Kpx_T \rho_o}{M_T f(t) + \alpha Kp M_T f(t) + Kpx_T \rho_o f(t) V_{gm}} \end{aligned} \tag{1.10}$$

In the formula,

$\rho_o$ crude oil density, kg/m$^3$;
$x_T$ mass fraction of ash, decimal;
$M_T$ the average relative molecular mass of the gas;
$V_{gm}$ molar volume, cm$^3$/mol;
$f(t)$ temperature coefficient, decimal.

To convert $C$ to volumetric concentration $C_v$ (cm$^3$/cm$^3$), simply multiply $C$ by 22,400. From the above formula, it can be seen that, based on the relationship between $K$ and α temperature, as well as the fundamental parameters of crude oil and gas, the equilibrium solubility of the gas can be calculated. Figure 1.4 shows the solubility coefficient of $CO_2$ in saturated crude oil sandstone.

In addition, the experimentally measured solubility $\rho_o$ and α values can be calibrated.

$$C_o = \frac{n_g}{V_o} = \frac{n_g V_2}{V_2 V_o} = CB = \frac{Kp\rho_o x_T}{(1+\alpha Kp) M_T} \tag{1.11}$$

In equation $C_o$—the solubility of gas in crude oil at the surface, expressed as the amount of dissolved gas per unit volume when two-phase equilibrium is reached under certain temperature and pressure conditions, in mol/cm$^3$.

Equation (1.11) indicates that when the solubility is $C_o$, the solubility equation takes a simple form. Moreover, since the volume of 1 mol of gas under standard conditions is 22,400 cm$^3$, a further Eq. (1.12) can be derived:

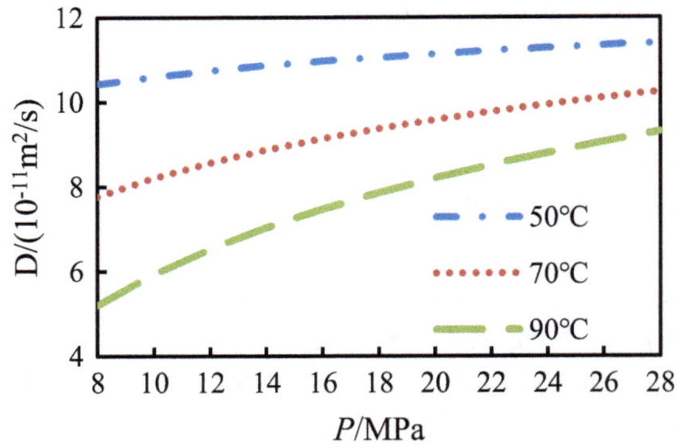


**Fig. 1.4** Solubility coefficient of $CO_2$ in saturated oil-bearing rocks

$$R_s = 22,400C_o = \frac{22,400Kpd_ox_T}{(1 + \alpha Kp)M_T} \tag{1.12}$$

In the formula, $R_s$—gas-oil ratio, $m^3/m^3$.

(2) Dissolution of $CO_2$ in Groundwater

The dissolution of $CO_2$ in formation water is one of the main forms of storage. Its solubility is influenced by formation temperature, formation pressure, and mineralisation. Considering only the above factors, it is usually expressed by Eq. (1.13):

$$M_w = \rho_r\left[C_{ws}\left(Ah\phi S_{wi} + V_{iw} - V_{pw}\right)\right] \tag{1.13}$$

In the formula,

$M_w$ theoretical $CO_2$ storage capacity in water, $10^6$ t;
$\rho_r$ density of $CO_2$ under formation conditions, $kg/m^3$;
$C_{ws}$ $CO_2$ solubility coefficient in water;
$A$ stratum area, $m^2$;
$h$ stratum thickness, m;
$\phi$ stratal porosity;
$S_{wi}$ stratum bundle water saturation;
$V_{iw}$ volume of water injected into the formation, $m^3$;
$V_{pw}$ the volume of water produced from the target formation, $m^3$.

Among these, the solubility in water is described using the activity coefficient model proposed by Huron, combined with the Peng–Robinson equation of state, to characterise the thermodynamic properties of the $CO_2$–$H_2O$ system. The equation of state is as follows:

$$\begin{cases} p = \frac{RT}{V-b_m} - \frac{a_m}{V(V+b_m)+b_m(V-b_m)} \\ a_m = b_m\left(\sum_{i=1}^{n} x_i \frac{a_i}{b_i}\alpha_i - \frac{G_\infty^E}{C_o}\right), b_m = \sum_{i=1}^{n} x_i b_i \\ \alpha_i = 0.477235\frac{R^2T_{ci}^2}{P_{ci}}\alpha_i, b_i = 0.07796\frac{RT_{ci}}{P_{ci}} \\ \alpha_i = \left[1 + m_i\left(1 - T_{ri}^{0.5}\right)\right]^2, c_o = \frac{1}{2\sqrt{2}}\ln\left(\frac{2+\sqrt{2}}{2-\sqrt{2}}\right) \end{cases} \quad (1.14)$$

In the formula,

$p$ pressure, MPa;
$R$ universal gas constant, 8.314 J/(mol K);
$T$ system temperature, K;
$V$ molar volume of the substance, $cm^3$/mol;
$a_m$ the gravitational constant of the composite system;
$b_m$ Poisson's ratio of the composite system;
$x_i$ the mass fraction of component $i$ in the liquid phase;
$a_i$ and $b_i$ the equation of state parameters for the pure substances of the $i$ components;
$G_\infty^E$ excess Helmholtz free energy under isochoric conditions, V/mol;
$T_{ci}$ $i$ component critical temperature, °C;
$P_{ci}$ $i$ component critical compressive strength, MPa;
$T_{ri}^{0.5}$ $i$ the comparative temperature of the components, $T_{ri}^{0.5} = \frac{T}{T_{ci}}$;
$m_i$ regarding the function of the eccentric factor $\omega$, $m = 0.37646+1.5426\omega-0.26992\omega^2$.

Under conditions of unlimited pressure, the expression proposed by Huron and Vidal is used to calculate the Gibbs free energy:

$$G_\infty^E = \sum_{i=1}^{n} x_i \frac{\sum_{j=1}^{n} x_j C_{ji} G_{ji}}{\sum_{k=1}^{n} x_k G_{ki}} \quad (1.15)$$

Among them, $C_{ji} = g_{ji} - g_{ii}$

$$\begin{cases} g_{ji} = -2\frac{b_i b_j}{b_i+b_j}\left(g_{ii}g_{jj}\right)^{0.5}\left(1 - k_{ij}\right), g_{ii} = -c_o\frac{a_i}{b_i} \\ G_{ji} = b_j \exp\left(-\alpha_{ji}\frac{C_{ji}}{RT}\right), G_{ki} = b_k \exp\left(-\alpha_{ki}\frac{C_{ki}}{RT}\right) \end{cases} \quad (1.16)$$

In the formula,

$g_{ji}$ the Lennard–Jones potential parameter for different intermolecular interactions;
$g_{ii}$ the Bjerrum length for identical intermolecular interactions;
$k_{ij}$ the interaction parameter between component $i$ and component $j$;
$G_{ji}$ and $G_{ki}$ temperature-dependent adjustable parameters;
$\alpha_{ji}$ the non-random parameter between components $i$ and $j$;
$\alpha_{ki}$ the non-random parameter between components $i$ and $k$.

Based on the principles of chemical thermodynamics, the selectivity coefficient of component $(T, p)$ with respect to the independent variable $i$ can be expressed as follows:

$$\boldsymbol{RT}\ln\varphi_i = \frac{\partial}{\partial n_i}\left[\int_{v_t}^{\infty}\left(p - \frac{n\boldsymbol{RT}}{v_t}\right)dv_t\right]_{T,v_t,n_i\neq j} - \boldsymbol{RT}\ln Z \tag{1.17}$$

In the formula,

| | |
|---|---|
| – | denotes the selectivity coefficients of the components; |
| $v_t$ | total system volume, $cm^3/mol$; |
| $n_i$ and $n_j$ | the mass of component $i$ and component $j$. |

The combination of the Peng–Robinson equation of state and the Huron-Vidal mixing rule yields the mixture fugacity coefficient $\varphi_m$:

$$\ln\varphi_m = \frac{b_i}{b_m}(Z_m - 1) - \ln\left[Z_m\left(1 - \frac{b_m}{v_m}\right)\right] - \frac{1}{2\sqrt{2}\boldsymbol{RT}}\left[\frac{a_i}{b_i} - \frac{\boldsymbol{RT}\ln\gamma_i}{c_o}\right] \ln\left[\frac{v_m + \left(\sqrt{2}+1\right)b_m}{v_m - \left(\sqrt{2}-1\right)b_m}\right] \tag{1.18}$$

In the formula,

| | |
|---|---|
| $\varphi_m$ | fugacity coefficient of the mixed material; |
| $Z_m$ | the deviation factor of the composite material; |
| $v_m$ | molar volume of the composite system, $m^3/mol$; |
| $\gamma_i$ | the activity coefficient of component $i$ in the mixture. |

Equations (1.19)–(1.21) are the calculation formulas for solubility values, while Fig. 1.5 presents the solubility coefficient diagram of $CO_2$ in saturated aquifer limestone. According to the state equation, it can be determined that the fugacity of $CO_2$ is equal in the two-phase system, thus:

$$f_{CO_2}^{v} = f_{CO_2}^{l} \tag{1.19}$$

The selectivity of $CO_2$ in any given pair can be determined by the selectivity coefficient as follows:

$$\begin{cases} f_{CO_2}^{v} = y_{CO_2}^{v}\varphi_{CO_2}^{v}p \\ f_{CO_2}^{l} = y_{CO_2}^{l}\varphi_{CO_2}^{l}p \end{cases} \tag{1.20}$$

Therefore, from Eqs. (1.19) and (1.20), the solubility of $CO_2$ in the liquid phase can be obtained as follows:

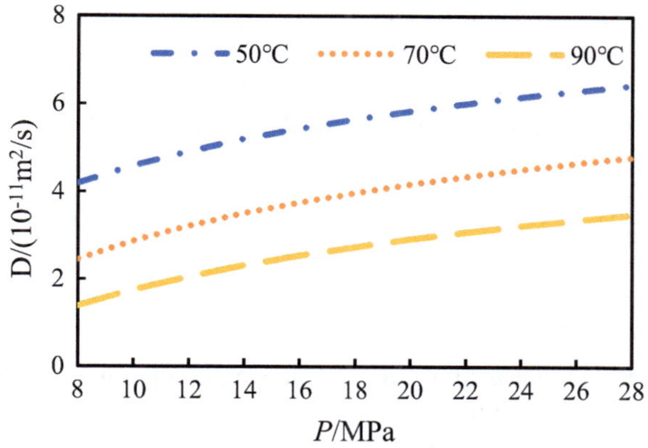


**Fig. 1.5** Solubility coefficient of $CO_2$ in saturated formation water

$$y_{CO_2}^{l} = \frac{\varphi_{CO_2}^{v}}{\varphi_{CO_2}^{l}} y_{CO_2}^{v} \tag{1.21}$$

(3) $CO_2$ Is Sequestered Within Effective Storage Pore Spaces

Assuming that the pore space within the effective storage zone of the reservoir is filled with $CO_2$, the formula for calculating the amount of $CO_2$ stored in the effective pore space of the reservoir is as follows:

$$M_w = \rho_r\left[E_R A h \phi (1 - S_{wi}) - V_{iw} + V_{pw}\right] \tag{1.22}$$

In the formula,

$M_w$ the theoretical storage capacity of $CO_2$ in the effective pore space, $10^6$ t;
$\rho_r$ density of $CO_2$ in the storage pore space, kg/m$^3$;
$A$ effective storage pore surface area, m$^2$;
$h$ storage space thickness, m;
$\phi$ storage pore spacing;
$S_{wi}$ storage pore bundle water saturation;
$V_{iw}$ volume of water injected into the storage space, m$^3$;
$V_{pw}$ the volume of water produced from the storage space, m$^3$.

(4) Chemical Capture of $CO_2$ by Minerals

According to experimental analysis, changes in the pore structure of sandstone are controlled by the initial porosity of the reservoir, the mineral composition of the sandstone, certain ions present in the acidic aqueous solution, and the active minerals within the sandstone. Ultimately, these factors result in alterations to the microstructure of the sandstone and changes in its porosity. As the contact time between the $CO_2$ aqueous solution and the sandstone increases, the dissolution process reduces the particle size, enlarges the contact area between the cementation surfaces and the chemical solution, and leads to a loosening of the framework grain structure of the sandstone.

According to the chemical equilibrium equation, the change in the mass of a given mineral involved in the chemical reaction with the reservoir rock can be expressed as:

$$\frac{\partial m}{\partial t} = \frac{\partial (n \cdot M)}{\partial t} = \frac{\partial (\rho V \cdot \phi)}{\partial t} \tag{1.23}$$

In the formula,

$m$ the mass of a given substance participating in a chemical reaction, g;
$t$ reaction time, s;
$\rho$ mineral density, g/cm$^3$;
$n$ the amount of mineral substance participating in the chemical reaction, mol;
$M$ the molar mass of a certain mineral component, mol/g;
$\phi$ porosity, %;
$V$ volume of the rock, cm$^3$.

As multiple minerals in the reservoir rock simultaneously undergo chemical reactions, the total volume reduction of all minerals participating in chemical reactions within the reservoir rock can be summed and expressed by Eq. (1.24):

$$\frac{\partial (V \cdot \phi)}{\partial t} = \sum_{i=1}^{\tau} \frac{\partial n_i}{\partial t} \cdot \frac{M_i}{\rho_i} \tag{1.24}$$

Chemical formula:

$$\frac{\partial \phi}{\partial t} = \frac{1}{V} \cdot \sum_{i=1}^{\tau} \frac{\partial n_i}{\partial t} \cdot \frac{M_i}{\rho_i} \tag{1.25}$$

## 1.2.3 Storage and Utilization Technology

Geological storage of $CO_2$ refers to the process of capturing and separating $CO_2$ from stationary carbon emission sources, compressing and liquefying it, and then injecting it underground for storage. The implementation of this process requires significant economic investment, which limits the practical application of $CO_2$ geological storage. However, $CO_2$-enhanced oil and gas recovery technologies can simultaneously sequester $CO_2$ underground while displacing oil and gas, thereby achieving both geological storage of $CO_2$ and improving the recovery rates of oil and gas. These technologies are known as $CO_2$-Enhanced Oil Recovery ($CO_2$-EOR), $CO_2$-Enhanced Coal Bed Methane recovery ($CO_2$-ECBM), and $CO_2$-Enhanced Gas Recovery ($CO_2$-EGR). This represents a "two birds with one stone" approach, significantly reducing

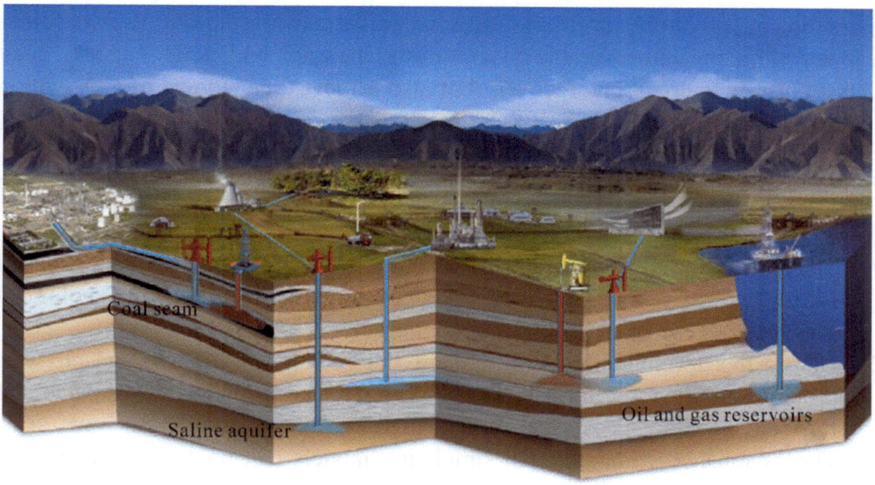


**Fig. 1.6** Principal settings for $CO_2$ geological storage

the cost of $CO_2$ storage. Moreover, implementing $CO_2$ storage projects in oil and gas fields that are currently under development or have already been depleted allows for the utilisation of existing regional geological data, such as exploration, drilling, and well-logging information, offering notable cost advantages. As a result, this is currently the main practical approach for $CO_2$ geological storage. Due to their high porosity, strong permeability, and widespread global distribution in sedimentary basins, saline aquifers possess the largest storage capacity and economic advantages, and thus saline aquifer storage technology is also developing rapidly. Overall, the main geological settings for $CO_2$ storage at present, as shown in Fig. 1.6, include deep coal seams, oil and gas reservoirs, and saline aquifers.

(1) $CO_2$-Enhanced Oil Recovery

Oil and gas reservoirs represent an excellent option for $CO_2$ storage and are currently among the most mature storage technologies available. There are generally two scenarios for $CO_2$ storage in oil and gas reservoirs. The first involves storage in depleted oil and gas fields, where $CO_2$ is directly injected into the reservoir in the form of a supercritical fluid. Oil and gas fields that have been sealed for extended periods typically exhibit excellent sealing properties, with hydrocarbons remaining trapped for tens of thousands or even millions of years without leakage. This demonstrates the high safety and potential of $CO_2$ storage in oil and gas reservoirs. Utilising existing oilfield infrastructure and technology also reduces both the cost of $CO_2$ storage and the need for new construction, while eliminating the necessity for further exploration and lowering labour costs. The second scenario involves $CO_2$-enhanced oil recovery ($CO_2$-EOR) technology. During the extraction of crude oil or natural gas, $CO_2$ is injected in the form of a supercritical fluid into producing oil and gas fields to enhance recovery rates. Under high-pressure conditions, $CO_2$ mobilises the remaining oil, promoting its flow towards the wellbore, while the $CO_2$ itself becomes trapped within the unrecoverable oil and subsurface water bodies. This approach not only improves hydrocarbon recovery but also achieves the storage of $CO_2$ within the oil and gas reservoir. Both of these methods are collectively referred to as $CO_2$ storage in oil and gas reservoirs.

Palaeoreservoirs are the preferred sites for $CO_2$ geological storage, with the following key advantages:

1) The fact that the crude oil within the structurally enclosed reservoir has not leaked for millions of years is sufficient to demonstrate the integrity of its geological structure and the safety of $CO_2$ containment.
2) Due to the development of oil storage, the geological structure and physical properties of the reservoir have been thoroughly understood, which is beneficial for reducing storage costs and enhancing safety.
3) The established computational simulation models can predict the migration, displacement mechanisms, and storage processes of subsurface $CO_2$.
4) Existing oil wells and field infrastructure are conducive to the implementation of $CO_2$ storage operations.
5) $CO_2$ does not affect the petroleum already present in depleted oil reservoirs, and if the oil field is still in production, $CO_2$ storage can also enhance oil recovery during the process.

$CO_2$-EOR technology refers to the process of injecting $CO_2$ into oil reservoirs, utilising its physical and chemical interactions with crude oil to alter the properties of both the oil and the reservoir. This industrial process aims to simultaneously enhance oil recovery and achieve $CO_2$ storage, making it one of the most promising directions within CCUS. $CO_2$ in a supercritical state exhibits strong solubility and extraction capabilities. After secondary recovery in oilfields, some crude oil remains trapped in the pore spaces of the reservoir rock due to capillary forces. Injecting $CO_2$ into the same formation can induce mechanisms such as crude oil swelling, viscosity reduction, improved oil–water mobility ratio, extraction of light components, and miscibility effects, thereby increasing oil recovery and boosting crude oil production.

Conventional primary oil recovery techniques are only able to extract 5–40% of the original oil in place from geological reserves. Secondary recovery methods, such as water flooding, can increase production by an additional 10–20% of the geological reserves. In contrast, tertiary oil recovery technologies, such as $CO_2$ injection for enhanced oil recovery, can further increase the recovery of crude oil by 7–23% (with an average of 13.2%) of the geological reserves.

Injecting fluids into underground oil and gas reservoirs to increase subsurface fluid pressure and enhance oil and gas recovery rates is an assisted recovery technology that has been widely applied in the oil and gas extraction industry. The use of $CO_2$ to replace conventional injection fluids for oil and gas displacement not only improves recovery rates but also reduces the cost of $CO_2$ geological storage. The injection of $CO_2$ can compensate for the decline in reservoir pressure caused by oil and gas extraction, displacing oil and gas within the pore spaces, thereby both increasing recovery rates and sequestering $CO_2$. The injected $CO_2$ partially mixes with the oil and gas fluids, increasing their pressure and reducing their viscosity, making oil and gas extraction easier. During the process of $CO_2$-enhanced oil (natural gas) recovery and storage, the majority of $CO_2$ is stored in subsurface structural and stratigraphic traps, a portion dissolves in the remaining subsurface fluids, a small fraction is mineralised and sequestered in underground rocks, and some $CO_2$ is produced back to the surface along with the oil and gas stream.

There are various methods for enhancing oil recovery using $CO_2$. For example, these include continuous $CO_2$ injection and alternating injection of $CO_2$ and water. The process by which injected $CO_2$ displaces crude oil is related to the phase behaviour characteristics of the $CO_2$-crude oil mixture, which in turn depend on the reservoir pressure, temperature, and the composition of the crude oil. In reservoirs with relatively low pressure, $CO_2$ injection enhances oil recovery primarily through mechanisms such as oil swelling and viscosity reduction; this process is referred to as immiscible displacement enhanced oil recovery technology (see Fig. 1.7). In contrast, in reservoirs with relatively high pressure, when the reservoir pressure exceeds the minimum miscibility pressure, the injected $CO_2$ becomes fully miscible with the crude oil in the reservoir, thereby achieving enhanced oil recovery; this process is known as miscible displacement enhanced oil recovery technology (see Fig. 1.8). During the application of $CO_2$ enhanced oil recovery, approximately 50–67% of the injected $CO_2$ is produced together with the crude oil. Generally, in order to reduce operational costs, the produced $CO_2$ is separated from the crude oil and reinjected into the reservoir for reuse. Ultimately, when oil production from the reservoir is no longer economically viable, the $CO_2$ will be permanently stored in the subsurface through various geological storage methods.

Generally, oil and gas reservoirs suitable for $CO_2$ geological storage are either depleted, having lost their economic production capacity, or are in the late stage of secondary development. After most oilfields have undergone secondary water injection, the pore spaces in the original oil and gas reservoirs are largely filled with formation water or injected water, resulting in the formation of aquifers. Therefore,

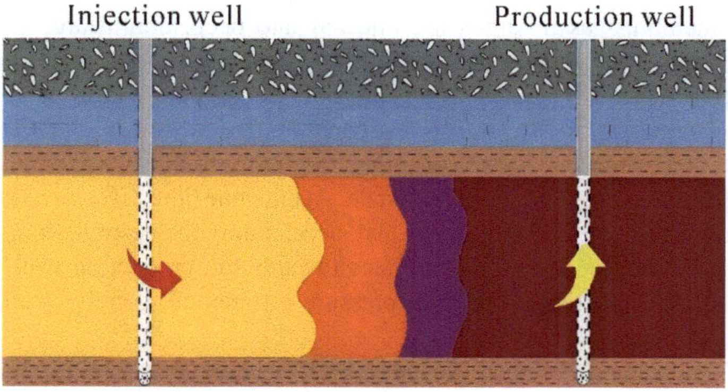


**Fig. 1.7** Schematic of $CO_2$ storage with enhanced oil recovery

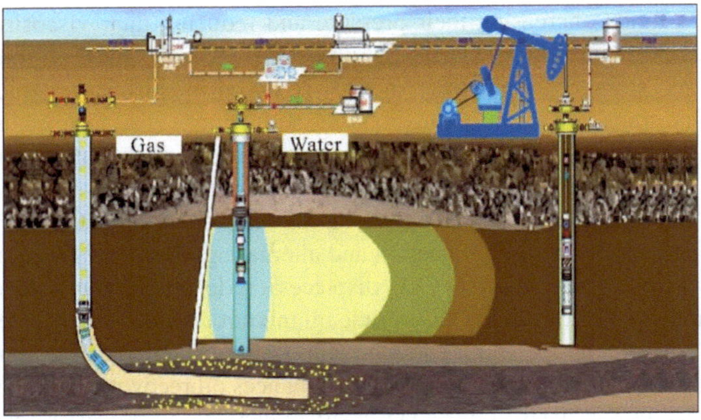


**Fig. 1.8** Schematic of $CO_2$-water miscible flooding for enhanced oil recovery and storage

the geological conditions for $CO_2$ storage in depleted oil and gas reservoirs are quite similar to those for $CO_2$ storage in deep saline aquifers.

As oil and gas reservoirs are sites where $CO_2$ has already been stored within existing reservoir formations, their storage capacity and safety have been validated, and they possess relatively high containment capability and robust security. This distinguishes them from deep saline aquifers used for $CO_2$ geological storage. Therefore, $CO_2$ storage in oil and gas reservoirs requires a renewed evaluation of reservoir sedimentary types (such as clastic or carbonate formations), burial depth, thickness, three-dimensional geometric morphology and integrity, as well as reservoir properties and heterogeneity, based on the foundational research conducted during oil and gas exploration and development. This enables an objective and realistic assessment of the $CO_2$ storage capacity.

(2) $CO_2$-Enhanced Coalbed Methane Recovery

$CO_2$-ECBM (Fig. 1.9) refers to the industrial process in which compressed $CO_2$ or $CO_2$-containing gas mixtures are injected into deep, unmineable coal seams. This process not only achieves long-term storage of $CO_2$ but also enhances coalbed methane (CBM) recovery. After $CO_2$ is injected into the coal seam, due to the coal's stronger adsorption capacity for $CO_2$ compared to $CH_4$, the injected $CO_2$ is preferentially adsorbed by the coal matrix, enabling long-term storage, while simultaneously displacing the adsorbed $CH_4$ into a free state. At the same time, $CO_2$ injection reduces the partial pressure of $CH_4$ within the coal, accelerating the desorption of $CH_4$. The liberated $CH_4$ then continuously migrates towards the production well under the influence of the injection-production pressure differential, thereby increasing the recovery rate of coalbed methane.

The $CO_2$ enhanced coalbed methane (ECBM) storage technology is primarily targeted at deep, unmineable coal seams. The roofs of such coal seams are typically composed of mudstone or shale and other non-permeable strata. By injecting $CO_2$ into the coal seam, the $CO_2$ displaces and replaces the methane gas stored within the

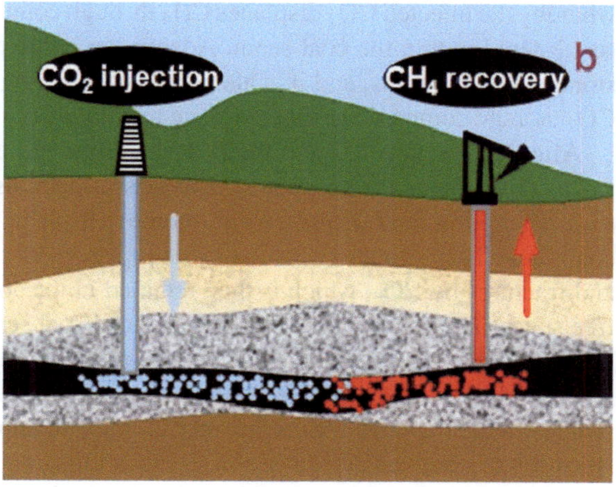


**Fig. 1.9** Schematic of $CO_2$-enhanced coalbed methane recovery

fractures or pores of the coal seam. This process enables the simultaneous storage of $CO_2$ and the enhanced recovery of coalbed methane. The economic benefits generated from coalbed methane extraction can, to a certain extent, offset the costs associated with $CO_2$ storage.

Coal seams, as geological reservoirs for $CO_2$, exhibit two notable characteristics: firstly, under certain temperature and pressure conditions, they possess the capacity to adsorb and store gases; secondly, coal seams are a typical dual-porosity medium, comprising two major systems—primary pores (such as micropores and intergranular pores) and secondary pores (such as macropores and natural fractures). The presence of macropores and fractures within coal seams enables the flow of gases. Therefore, pores serve as the primary storage sites for $CO_2$ in coal seams, while fractures act as conduits connecting pores to each other and linking pores with fractures, thus providing the main pathways for $CO_2$ migration within the coal seam.

The free and adsorbed coalbed gas present in coal seams is the dynamic equilibrium product formed under confined conditions, following multiple cycles of adsorption–desorption, diffusion, permeation, and migration during the processes of coalification, structural activity, and burial evolution. The development of coalbed methane essentially involves the kinetic process of gas desorption from the coal seam. From the perspective of surface extraction of coalbed methane, the process can be summarised as a continuous sequence of desorption, diffusion, and permeation. In fact, the geological storage of $CO_2$ in coal seams can be simplified as the reverse process of coalbed methane extraction, with the core mechanism being the kinetic process of $CO_2$ adsorption and its displacement of $CH_4$. Therefore, the mechanism of $CO_2$ geological storage in unmineable coal seams is fundamentally governed by the adsorption–desorption behaviour of $CO_2$ within the pore-fracture structure of the coal seam.

The primary component of coalbed gas is $CH_4$. The adsorption capacity of coal seams for $CO_2$ is approximately twice that for $CH_4$. The $CO_2$-ECBM technology exploits the competitive adsorption relationship between $CO_2$ and $CH_4$ in coal seams, whereby the injected $CO_2$ displaces $CH_4$ through competitive adsorption while being sequestered within the coal seam. Adsorption is the principal mechanism for $CO_2$ storage in coal seams, and it is also one of the distinguishing features of coal seam $CO_2$ storage compared to other storage technologies.

After $CO_2$ is injected into the coal seam, the mixed fluids within the pores and fractures of the coal seam undergo processes such as diffusion, seepage, competitive adsorption, and displacement, thereby displacing the $CH_4$ gas present in the coal seam. This ultimately results in the replacement of $CH_4$ molecules in the pores and fractures by $CO_2$, which is then retained in the coal seam in both adsorbed and free states. The technology for $CO_2$ enhanced coal seam gas storage is not yet fully mature, with several technical challenges remaining to be addressed and requiring further research. For example, the injection of $CO_2$ may lead to a reduction in coal seam permeability, making continued $CO_2$ injection impossible. In addition, the classification of unmineable coal seams is subject to economic and technical constraints at specific periods; with economic and technological development, unmineable coal seams may become mineable. However, once $CO_2$ has been injected, it will be difficult to reopen and exploit these seams in the future.

(3) $CO_2$-Enhanced Natural Gas Recovery

$CO_2$-enhanced natural gas recovery ($CO_2$-EGR) technology refers to the process of injecting $CO_2$ into the depleted zones at the bottom of natural gas reservoirs, thereby displacing the residual natural gas that cannot be extracted due to natural decline, while simultaneously storing the $CO_2$ within the geological structure of the reservoir to achieve $CO_2$ emission reduction. On one hand, the injection of $CO_2$ is beneficial for restoring reservoir pressure; on the other hand, the high density and low viscosity characteristics of supercritical $CO_2$ are advantageous for creating gravitational segregation with natural gas. Under the influence of gravitational segregation, supercritical $CO_2$ tends to sink towards the lower part of the reservoir, forming a "gas cap" of natural gas, which promotes the upward migration and production of natural gas. Additionally, due to the weak convective mixing between $CO_2$ and natural gas in the reservoir, the two can maintain a non-equilibrium state of incomplete mutual dissolution over a relatively long period.

Based on the proven natural gas geological reserves, the $CO_2$ storage capacity of major basins in China is 30.483 billion tonnes, of which approximately 78.1% is located in onshore areas and about 21.9% in offshore continental shelf regions.

(4) $CO_2$ Storage in Aquifers

$CO_2$ saline aquifer storage (Fig. 1.10) involves injecting high-density $CO_2$ under pressure into saline aquifers via injection wells, where it fills the pore spaces, partially displacing the original brine. This process enables the safe, stable, and long-term industrial storage of $CO_2$. Globally, underground saline aquifers are widely distributed and possess enormous storage potential, making them one of the most viable technical options for $CO_2$ storage.

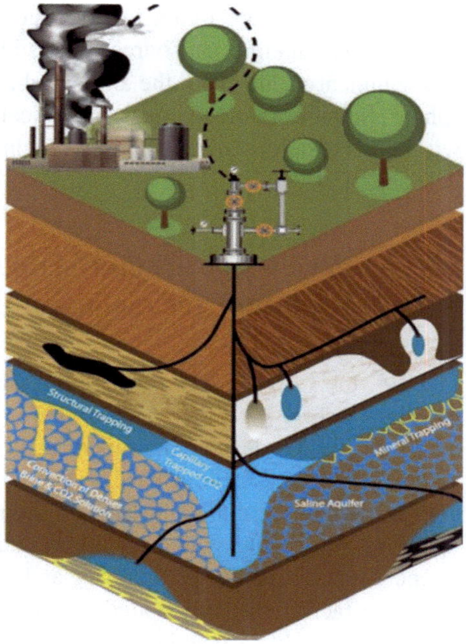


**Fig. 1.10** Schematic of $CO_2$ storage in saline aquifers

The geological storage mechanisms of $CO_2$ can be categorised into two main types: physical trapping and chemical trapping. Physical trapping includes structural and stratigraphic trapping, residual gas trapping, and hydrodynamic trapping; chemical trapping encompasses solubility trapping and mineral trapping. In the initial stage of $CO_2$ injection, structural and stratigraphic traps play the dominant role. Over time, the amount of mobile $CO_2$ within structural and hydrodynamic traps gradually decreases, and is progressively replaced by residual gas trapping and solubility trapping, while mineral trapping mechanisms also begin to take effect. Furthermore, even if the above mechanisms fail to capture $CO_2$, under certain favourable conditions, the fluid flow rate within the formation is relatively low, and it may take millions of years for $CO_2$ to migrate to the surface. A large quantity of $CO_2$ can thus be effectively sequestered by this method, which is referred to as the fluid dynamic trapping mechanism.

The technology of $CO_2$ storage in aquifers fully exploits the supercritical properties of $CO_2$. Under supercritical conditions (i.e., temperature not lower than 31.1 °C and pressure not less than 7.38 MPa), $CO_2$ transforms into a supercritical fluid, which possesses a high density similar to that of a liquid (approximately 80–400 times the density under standard atmospheric pressure) and the mobility characteristic of a gas. This enables $CO_2$ to migrate rapidly and in large quantities within geological formations, occupying the pore spaces or fractures within the strata. Driven by buoyancy, supercritical $CO_2$ accumulates at the base of the caprock, gradually filling the entire pore space of the reservoir. A portion of the $CO_2$ dissolves in formation water and reacts with ions and minerals, ultimately achieving long-term storage.

$CO_2$ saline aquifer storage involves the injection of $CO_2$, primarily captured from major industrial emission sources, into suitable saline aquifers through engineering and technical means, with the aim of achieving long-term isolation from the atmosphere. Suitable saline aquifers are typically located at depths greater than 800 m, with mineralisation levels generally ranging from 3 to 50 g/L.

## 1.3 Overview of Carbon Dioxide Geological Storage Projects and Technological Development

In 2025, the $CO_2$ geological storage industry experienced unprecedented growth, as shown in Fig. 1.11, with a total of 734 projects under development. The number of projects under development increased by 17% compared to 2024, reflecting the growing global recognition that $CO_2$ geological storage is a crucial tool for addressing climate change. In addition, there are currently 77 $CO_2$ geological storage facilities in operation worldwide, with 47 facilities under construction. At present, the facilities in operation are capable of capturing and storing 51 million tonnes of $CO_2$ annually. Once the facilities currently under construction become operational, this figure is expected to double, exceeding 100 million tonnes per year.

The United States is at the forefront globally in the utilisation of $CO_2$ resources and geological storage. As early as the 1970s, the US initiated research related to the use of $CO_2$ to enhance oil recovery. In order to achieve large-scale reductions

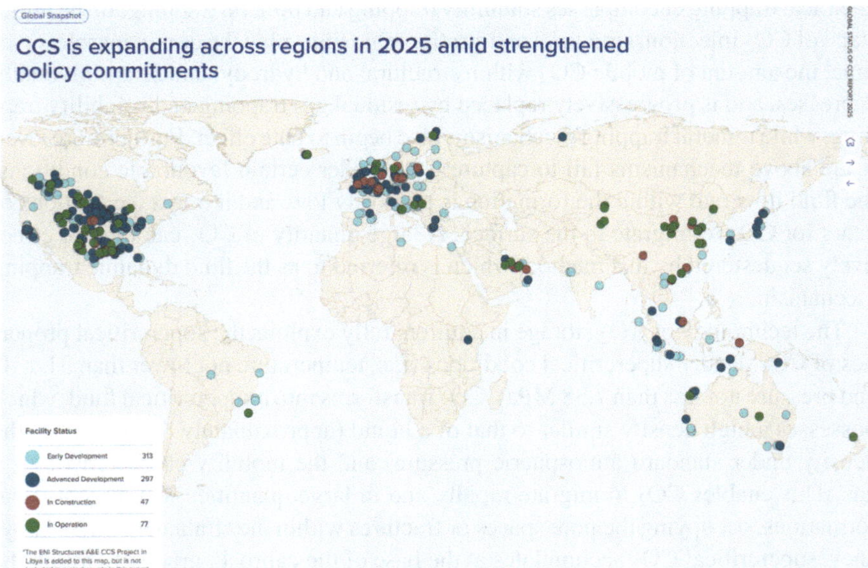


Fig. 1.11 Global status of $CO_2$ geological storage projects (GLOBAL STATUS OF CCS 2025)

in greenhouse gas emissions, the European Union launched the GESTCO project, which aimed to assess the potential for geological storage of $CO_2$ emitted from the combustion of fossil fuels in Europe. The project sought to evaluate the geological storage capacity for $CO_2$ in Europe, propose rational and scientific concepts for carbon storage, and ensure a stable energy supply for the continent. Norway's national oil company commenced the Sleipner carbon storage project in the North Sea in 1996. This remains the world's most successful $CO_2$ storage project in saline aquifers to date, and is also the first commercial project dedicated to geological storage of $CO_2$. No leakage has been detected in this project, demonstrating that, from both a technical and engineering perspective, the deep saline aquifer storage of $CO_2$ is feasible. In September 2000, the International Atomic Energy Agency launched a $CO_2$ geological storage project at the Weyburn oil field in Saskatchewan, Canada. After ten years of operation, approximately 18 million tonnes of $CO_2$ had been stored in the Weyburn oil field, with $CO_2$ injection also being used to enhance oil recovery. The storage project is expected to continue until 2035. In 2001, the European Union initiated the RECOPOL project in Poland, which was the first European pilot demonstration project for $CO_2$ storage in coal seams and for enhancing coalbed methane recovery. At present, research is ongoing into the migration and monitoring of $CO_2$ following injection. In 2004, the In Salah geological storage project in central Algeria began injecting $CO_2$ into a depleted gas field, achieving $CO_2$ storage in a depleted gas reservoir. As the world's first industrial-scale storage project, it also conducted extensive geophysical monitoring, with InSAR data providing real-time feedback on surface deformation during the $CO_2$ injection and storage process. Other countries have successively carried out assessments of $CO_2$ storage capacity in coal basins. In Alberta, Canada, the total $CO_2$ storage capacity in coal seams has reached $1.0 \times 10^{10}$ tonnes. In the Netherlands, field trials of ECBM technology and $CO_2$ geological storage have been conducted, and the $CO_2$ storage potential of four coalbed methane production areas, including Achterhoek, has been evaluated. The results indicate that $5.4 \times 10^7$–$9.0 \times 10^9$ tonnes of $CO_2$ can be stored in coal seams annually. Faiz has assessed the $CO_2$ storage capacity of bituminous coal in the southern Sydney Basin, Australia.

$CO_2$ geological storage projects have also been successively launched in China, resulting in the establishment of several $CO_2$ geological storage demonstration projects. The Shenhua $CO_2$ geological storage project, located in the deep saline aquifers of Yijinhuoluo Banner, Ordos City, Inner Mongolia, is the first full-process, coal-based deep saline aquifer $CO_2$ geological storage demonstration project in China. On 29 January 2022, Sinopec commenced operation of China's first million-tonne-scale carbon storage project—the Qilu Petrochemical–Shengli Oilfield Carbon Storage Project—which has now been fully constructed. The project is expected to reduce $CO_2$ emissions by one million tonnes annually. In June 2023, China National Offshore Oil Corporation (CNOOC) launched the country's first offshore $CO_2$ geological storage project, located in the eastern South China Sea oilfields. Within six months of its initiation, a total of 20 million $m^3$ of $CO_2$ had been injected and stored. The Ningxia 3 million tonnes/year $CO_2$ storage demonstration project has been launched at the Ningdong Energy and Chemical Base. Once operational, this project will become the world's first large-scale integration of gas injection for enhanced oil and gas recovery with modern coal chemical industry. During China's 13th Five-Year Plan, key projects were initiated to conduct fundamental research and techno-economic assessments on $CO_2$-enhanced coalbed methane recovery ($CO_2$-ECBM). As a major developing country, China has summarised and learnt from international achievements and experiences, and, in combination with domestic circumstances, has adopted a government-led, enterprise-driven model to carry out $CO_2$ geological storage projects. At the same time, China has actively sought international cooperation in the field of $CO_2$ geological storage. For example, from October 2003 to August 2004, the Chinese and Canadian governments jointly conducted pilot studies on $CO_2$ storage and enhanced coalbed methane recovery technologies at Well TL-003 in the southern part of the Suihua Basin, with a total injection of 192.8 tonnes of liquid $CO_2$, of which approximately 82% was ultimately sequestered in the coal seam. In 2018, the China Geological Survey led a comprehensive geological survey project on $CO_2$ geological storage in the Junggar Basin, Xinjiang, among others.

Hundreds of $CO_2$ geological storage projects have been carried out worldwide, as shown in Table 1.1, which presents several engineering case studies.

## 1.4 Section Summary

This chapter focuses on issues related to geological storage of $CO_2$, providing a detailed introduction to global climate change and corresponding mitigation measures. It analyses the variations in atmospheric $CO_2$ concentrations and their impacts on ecosystems, and outlines the measures adopted by various countries worldwide in response to climate change, with particular emphasis on the emission reduction strategies proposed by China to overcome significant challenges during its economic development.

Geological storage of $CO_2$ is a key measure for achieving carbon neutrality. Structural trapping, residual trapping, solubility trapping, and mineral trapping are the four main mechanisms of $CO_2$ geological storage. Sealed and stable storage spaces constitute the sites for $CO_2$ geological storage, and the storage capacity varies significantly under different trapping mechanisms. Therefore, this section provides a detailed overview of the calculation methods for storage capacity under various trapping scenarios. At the current stage of development, $CO_2$-enhanced oil recovery, $CO_2$-enhanced coalbed methane recovery, $CO_2$-enhanced natural gas recovery, and $CO_2$ saline aquifer storage are the main approaches for $CO_2$ storage and utilisation.

Finally, from a global perspective, the implementation status of $CO_2$ geological storage projects worldwide is summarised, and the development of technologies related to $CO_2$ geological storage is analysed.

**Table 1.1** Selected engineering case studies of $CO_2$ geological storage projects

| Project name | Country | Storage reservoir | Period | Scale |
|---|---|---|---|---|
| Shizhuang South CBM Block ECBM project | China | Coal reservoir | 2003–2004 | A total of 192.8 t of liquid $CO_2$ was injected, and approximately 82% of the $CO_2$ was ultimately sequestered in the coal reservoir |
| Zama project | Canada | Oil reservoir/ EOR | 2006–present | Expected to operate for 18 years, with $1.3 \times 10^6$ t of $CO_2$ sequestered |
| Snohvit project | Norway | Saline formation | 2008–present | $CO_2$ is injected at $0.7 \times 10^6$ t/a; $3 \times 10^6$ t of $CO_2$ has been successfully sequestered to date |
| Otway Basin project | Australia | Depleted gas reservoir | 2008–2011 | Injection rate of 150 t/d, with a total of $0.65 \times 10^6$ t of $CO_2$ sequestered |
| Cranfield project | United States | Saline formation | 2009–2015 | Injection rate of $1.5 \times 10^6$ t/a, with a total of $5.37 \times 10^6$ t of $CO_2$ sequestered |
| Shizhuang North CBM Block ECBM project | China | Coal reservoir | 2010 | A total of 233.6 t of $CO_2$ was injected into Well SX-001 over 17 days of operation |
| Ordos project | China | Saline formation | 2011–present | $CO_2$ is injected at $0.1 \times 10^6$ t/a; by August 2012, $4 \times 10^4$ t of $CO_2$ had been successfully sequestered |
| Decatur project | United States | Saline formation | 2011–2014 | $CO_2$ was injected at 1100 t/a, with a total of $1 \times 10^6$ t of $CO_2$ sequestered |
| Port Arthur project | United States | Oil reservoir/ EOR | 2013–present | $CO_2$ is injected at $1 \times 10^6$ t/a; by early May 2013, more than $0.2 \times 10^6$ t of $CO_2$ had been successfully sequestered |
| Northern ReefTrend project | United States | Depleted oil reservoir | 2013–present | $CO_2$ is injected at 1000 t/d; by the end of 2015, more than $0.46 \times 10^6$ t of $CO_2$ had been successfully sequestered |
| Alberta Carbon TrunkLine project | Canada | Oil reservoir/ EOR | 2017–present | $CO_2$ is injected at $14.6 \times 10^6$ t/a, with expected $CO_2$ storage exceeding $20 \times 10^8$ t |

## References


1. Zhang L, Li R (2020) Geological utilisation and storage of carbon dioxide: risk management. Science Press, Beijing
2. Li Q (2021) Discussion on China's energy transition pathways under the carbon neutrality target. China Coal 47(08):1–7
3. Peng S, Shi S, Wang J (2025) Geological storage and monitoring of carbon dioxide. Machinery Industry Press, Beijing
4. Cai Y, Wang H, Zheng H et al (2021) Climate change tipping points and responses—carbon neutrality. China Popul Resour Environ 31(09):16–23
5. Zhao XL, Liao XW, Wang WF et al (2014) The $CO_2$ storage capacity evaluation: methodology and determination of key factors. J Energy Inst 87(4):297–305
6. Lei Z, Li Z (2023) Fundamentals of geological storage and utilisation monitoring of carbon dioxide. China Petrochemical Press, Beijing
7. Xu Y (2025) Mechanisms and potential of geological storage of $CO_2$ in the Xinsheng generation of the Zhuwu Formation, Zhuji County, Zhejiang Province. Jilin University
8. Zhao X (2025) Numerical simulation study on the geological storage patterns of carbon dioxide in typical coal seams of the Huainan Coalfield. Anhui University of Science and Technology
9. Anderson TR, Hawkins E, Jones PD (2016) $CO_2$, the greenhouse effect and global warming: from the pioneering work of Arrhenius and Callendar to today's earth system models. Endeavour 40(3):178–187
10. Cramer W, Bondeau A, Woodward FI et al (2001) Global response of terrestrial ecosystem structure and function to $CO_2$ and climate change: results from six dynamic global vegetation models. Glob Change Biol 7(4):357–373
11. Rogelj J, Den Elzen M, Hohne N et al (2016) Paris Agreement climate proposals need a boost to keep warming well below 2°C. Nature 534:631–639
12. Bodansky D (2017) The Paris climate change agreement: a new hope. Am J Int Law 110(2):288–319
13. Xie Z (2021) Persistently and actively responding to climate change: continuing to be an important participant, contributor, and leader in global ecological civilisation construction—commemorating the fifth anniversary of the Paris Agreement. Environ Sustain Dev 46(1):3–10
14. Wang F, Harindintwali JD, Yuan ZZ et al (2022) Technologies and perspectives for achieving carbon neutrality. Innovation 2(4):100180
15. Li D (2021) Study on fine description and migration monitoring methods of $CO_2$ storage reservoirs. China University of Mining and Technology, Beijing
16. Rong Y, Xu P (2018) Progress in research on geological storage and utilisation technologies for $CO_2$. China Basic Sci 20(4):44–48
17. Michael K, Golab A, Shulakova V et al (2010) Geological storage of $CO_2$ in saline aquifers—a review of the experience from existing storage operations. Int J Greenh Gas Control 4(4):659–667
18. Lei Y (2021) Annual report on carbon capture, utilisation and storage (CCUS) in China (2021): recommendations for large-scale CCUS demonstration and industrial cluster construction. Environ Econ 16:40–42
19. Talebian SH, Masoudi R, Tan IM et al (2014) Foam assisted $CO_2$-EOR: a review of concept, challenges, and future prospects. J Petrol Sci Eng 120:202–215
20. Zhao CZ, Lu D, Chen K et al (2021) Review of density measurements and predictions of $CO_2$-alkane solutions for enhancing oil recovery. Energy Fuels 35(4):2914–2935
21. Yao C, Wu S, Yang Z et al (2023) Progress, challenges and significance of building a carbon industry system under the background of carbon neutrality strategy. Petrol Explor Dev 50(1):190–205
22. Qin J, Li Y, Wu D et al (2020) Global progress of CCUS and policy recommendations for China. Oil Gas Geol Recov Effic 27(1):20–28
23. Liu SY, Song YC, Zhao CZ et al (2018) The horizontal dispersion properties of $CO_2$-$CH_4$ in sand packs with $CO_2$ displacing the simulated natural gas. J Nat Gas Sci Eng 50:293–300

24. Reid TB, Robinson HJ (1981) Lick creek meakin sand unit immiscible $CO_2$ waterflood project. J Petrol Technol 33(9):1723–1729
25. Jiang H, Shen P, Chen L et al (2007) Current status of enhanced oil recovery by carbon dioxide in the North American petroleum industry. China Energy 29(7):30–34
26. Moritis G (2006) $CO_2$ injection gains momentum. Oil Gas J 104(15):37–41
27. Qian B, Zhu J (2008) Current status and prospects of $CO_2$ flooding for enhanced oil recovery worldwide. Energy Environ Prot 22(1):2–5
28. Koottungal L (2014) 2014 worldwide EOR survey. Oil Gas J 112(4):79–91
29. Qin J, Han H, Liu X (2015) Application and enlightenment of carbon dioxide flooding in the United States of America. Petrol Explor Dev 42(2):232–240
30. Yamaguchi S, Ohga K, Fujioka M et al (2006) Field test and history matching of the $CO_2$ sequestration project in coal seams in Japan. Int J Soc Mater Eng Resour 13(2):64–69
31. Liu Y, Li X, Bai J (2005) Preliminary evaluation of $CO_2$ storage capacity in Chinese coal seams. J Rock Mech Eng 24(16):2947–2952

# Chapter 2
# Theory and Mechanisms of Carbon Dioxide Geological Storage

The theoretical framework and mechanistic system of $CO_2$ geological storage constitute the core scientific foundation for understanding the essence of the storage process and guiding engineering practice. This section begins with an analysis of the fundamental conditions, providing a detailed exposition of the physicochemical properties of $CO_2$, the characteristics of storage media, and the geological compatibility requirements, thereby offering material and environmental bases for feasibility assessment. It then delves into the core theoretical models, including fluid substitution theory (incorporating the Gassmann equation and the variation patterns of reservoir elastic parameters) and multiphase-multifield coupling theory (covering seepage, convective dispersion, rock deformation, temperature fields, and their interrelations), establishing quantitative analytical and predictive tools for the storage process. Finally, it focuses on typical scenario mechanisms, elucidating the intrinsic mechanisms of $CO_2$-enhanced oil recovery (EOR), enhanced coalbed methane recovery (ECBM), and saline aquifer storage. The content of this section not only consolidates the theoretical foundation of $CO_2$ geological storage, but also provides key scientific support for engineering scheme design, optimisation of storage effectiveness, and long-term safety assessment, serving as a crucial link between fundamental research and practical application.

## 2.1 Physical Properties and Geological Conditions of Carbon Dioxide Storage Body

The physical properties of the storage medium and the geological foundation are the primary determinants for $CO_2$ geological storage. This section first analyses the physicochemical properties of $CO_2$, elucidating its phase transitions and migration behaviour in the subsurface environment; it then describes the material characteristics of storage media, examining the material basis for interactions between reservoir



D. Li and Y. Guo, *Geological Storage and Monitoring of Carbon Dioxide*,
https://doi.org/10.1007/978-981-92-3621-3_2

matrices and $CO_2$; finally, it systematically reviews the geological conditions suitable for storage, clarifying the key environmental factors for site selection.

### 2.1.1 *Physicochemical Properties of Carbon Dioxide*

The migration of $CO_2$ in porous media is accompanied by continuous phase changes due to variations in environmental pressure and temperature. As its physical properties change, these phase transitions exert a significant influence on the rate of percolation, thereby affecting the entire transport process. Therefore, analysing the physicochemical properties of $CO_2$ and the impact of phase transitions is of critical importance for $CO_2$ geological storage.

(1) Phase Characteristics

Under the influence of temperature and pressure, $CO_2$ can transition from a gaseous state to a liquid or solid state. As shown in Fig. 2.1, which presents the phase diagram of $CO_2$, the critical point occurs at a temperature of 31.1 °C and a pressure of 7.38 MPa. When both temperature and pressure continue to increase beyond this point, $CO_2$ enters a supercritical state. In this state, $CO_2$ exhibits a density close to that of a liquid and a viscosity similar to that of a gas, resulting in favourable properties such as high fluidity, high density, and strong solvating ability. Therefore, in $CO_2$ storage, injection is typically carried out under supercritical conditions, which is advantageous for the migration and large-scale storage of $CO_2$.

**Fig. 2.1** Phase diagram of $CO_2$

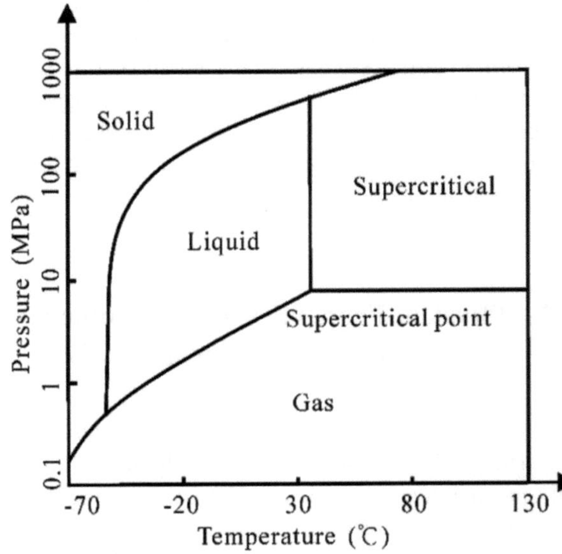

(2) $CO_2$ Concentration

When $CO_2$ is in the gaseous state, its density is greatly influenced by temperature and pressure; it increases with rising pressure and decreases with increasing temperature. Generally, the density of $CO_2$ is greater than that of natural gas at the gas cap, but less than that of crude oil. When $CO_2$ is in the liquid state, its density decreases with increasing temperature, but is less sensitive to changes in pressure. Under specific temperature $T_1$ and pressure $P_1$, the density can be calculated from the density at temperature T and pressure P, as shown in Eq. (2.1):

$$\rho_1 = \frac{\omega}{\omega_1} * \rho \tag{2.1}$$

In the equation, $\omega$ and $\omega_1$ represent the compressibility coefficients under the two respective states.

When $CO_2$ is in the solid state, its density is scarcely affected by temperature and pressure, or the effect is minimal. As shown by the Shengjinchang model, under constant pressure, the density of $CO_2$ decreases slightly with increasing temperature; under constant temperature, the density increases slightly with increasing pressure, and the rate of increase is related to the pressure.

(3) $CO_2$ Viscosity

When $CO_2$ is in the gaseous state, its viscosity is a function of temperature and pressure, as shown in Eq. (2.2):

$$\mu = a \times \sqrt{\frac{MT}{d^2\Omega_\mu}} \tag{2.2}$$

In the equation, $\mu$ denotes the viscosity of gaseous $CO_2$; $a = 2.6693 \times 10^{-3}$ is a constant; $M$ represents the molar mass; $T$ is the absolute temperature; $d$ denotes the effective molecular diameter; and $\Omega_\mu$ is a regression constant, calculated based on empirical models.

The viscosity of supercritical $CO_2$ exhibits a nonlinear relationship with temperature and a positive correlation with pressure. At lower pressures, viscosity increases with rising temperature, whereas at higher pressures, viscosity decreases as temperature increases.

The $CO_2$ content in fluid can be calculated using the empirical formula (2.3):

$$\eta = \frac{0.399}{V} e^{3.83\frac{T_b}{T}} \tag{2.3}$$

In the equation, $\eta$ denotes the viscosity of liquid $CO_2$; $V$ represents the molar volume of the liquid; $T_b$ is the boiling point of $CO_2$ at 0.1 MPa; and $T$ indicates the absolute temperature.

## 2.1.2 Physical Properties of the Carbon Dioxide Storage Body

(1) Characteristics of Porosity and Permeability

Porosity and permeability are two critical parameters of geological storage reservoirs for $CO_2$. Porosity determines the storage capacity of the reservoir for $CO_2$, while permeability determines the ability of the rock to allow fluids to pass through under a certain pressure. It is generally accepted that geological formations with relatively high porosity and permeability are suitable for $CO_2$ storage, which is crucial for the storage of large quantities of $CO_2$. Porosity refers to the portion of an aquifer composed of void spaces; however, not all pores within the reservoir contribute to $CO_2$ storage. The effective porosity, formed by the interconnected pores, determines the effective water storage capacity of the aquifer. The amount of $CO_2$ that can be stored in a given saline aquifer can be calculated using the volumetric factor $\eta$ as follows:

$$C = C_{gas} + C_{liquid} \tag{2.4}$$

where

- $C_{\text{gas}}$ the volumetric fraction (%) of mobile $CO_2$ in the reservoir, $C_{\text{gas}} = \langle \varphi \cdot S_g \rangle$;
- $C_{\text{liquid}}$ volume fraction (%) of $CO_2$ dissolved in the aquifer within the reservoir, $C_{\text{liquid}} = \langle \varphi \cdot S_L \cdot X_L^{gas} \cdot \rho_L/\rho_g \rangle$;
- $C$ the volumetric fraction (%) of sealable pore space within the reservoir, which is the sum of $C_{\text{gas}}$ and $C_{\text{liquid}}$;
- $\varphi$ effective porosity (%);
- $S_L$ volume fraction (%) of liquid-phase pore space;
- $S_g$ $CO_2$ gas-phase pore space volume fraction (%);
- $X_L^{gas}$ mass fraction (%) of dissolved $CO_2$ in brine;
- $\rho_g$ $CO_2$ density;
- $\rho_L$ liquid phase density;
- <> average value of porosity in the reservoir.

Effective porosity $\varphi$ plays a crucial role in determining the $CO_2$ storage capacity in aquifers and can serve as an important parameter for assessing the suitability of saline aquifers for storage. The volume of original fluids in the reservoir displaced by $CO_2$ ($V_{bulk}$) is a function of the injection volume ($V_{CO2}$) and the average porosity of the saline aquifer, namely:

$$V_{bulk} = \frac{V_{CO2}}{\varphi}\frac{1}{\psi} \tag{2.5}$$

In the formula, $\psi$—the $CO_2$ flooding efficiency coefficient—represents the pore and permeability characteristics of the reservoir and the gas injection process.

Essentially, the effective porosity of aquifers is a key factor determining their $CO_2$ storage capacity. Its value decreases with increasing effective stress; therefore, the porosity of the reservoir diminishes with depth. The size and structure of aquifer pores also vary under the same influences. Due to the considerable complexity of the pore structure in any porous medium—in terms of pore number, size, shape, orientation, and connectivity—the consideration of pore size distribution in aquifers often becomes particularly important. The pore size distribution is defined as the distribution function within the total pore size range of $(\sigma, \sigma + d\sigma)$, that is,

$$\int_0^\infty f(\sigma)d\sigma = 1 \tag{2.6}$$

Information on pore size distribution is crucial for characterising the displacement of brine by $CO_2$ in saline aquifers, as it can be used to calculate relative permeability. Following $CO_2$ injection, physicochemical changes within the reservoir, such as mineral precipitation, dissolution, and changes in effective stress, are factors that drive the evolution of pore size distribution. Mineral precipitation in porous media is often influenced by the size of the pores. While larger pores allow for mineral precipitation, smaller pores may inhibit this process, resulting in an overall reduction in reaction rates, which in turn stabilises porosity. Mineral trapping and dissolution trapping are key mechanisms for long-term storage, and the heterogeneous distribution of pore sizes may affect the ease with which reactions occur in different parts of the precipitate, as well as the dissolution rate of $CO_2$. Permeability is commonly expressed in Darcy units, defined as the volume of fluid passing through a unit cross-sectional area of rock per unit time under a unit pressure gradient. Darcy's law can be expressed as follows:

$$Q = -k \cdot A \cdot \frac{\Delta P}{\mu \Delta L} \tag{2.7}$$

In the form

$Q$ via the core flow rate ($cm^3/s$);
$k$ permeability of the rock ($m^3$);
$A$ cross-sectional area of the rock ($cm^2$);
$\Delta P$ pressure difference (MPa) in the upward direction of fluid flow;
$\Delta P$ viscosity of the fluid (Pa s);
$\Delta L$ length of the core (cm).

The value of permeability is determined by the pore structure characteristics of the porous medium. The target reservoirs for $CO_2$ storage must maintain an appropriate level of permeability to ensure effective storage of $CO_2$. The permeability of the reservoir is influenced by reactions involving rock minerals, as well as by dissolution and precipitation processes. Modelling indicates that, following $CO_2$ injection, the initial dissolution of carbonate cement increases the porosity of the precipitates; however, subsequent reactions lead to the dissolution of feldspar and the precipitation of carbonate minerals and clay, thereby reducing both permeability and porosity.

This suggests that the original permeability of the precipitates may change during the injection process, which will affect predictions of $CO_2$ migration.

Complex multiphase flow-solid systems exist within underground reservoirs, and the process of $CO_2$ storage is by no means a simple single-phase flow. Therefore, it is necessary to introduce relative permeability in order to study the two-phase migration relationship between $CO_2$ and brine systems within the reservoir. Permeability is an intrinsic property of porous media, while relative permeability is a characteristic of the fluid–fluid–solid system. Although relative permeability is determined by many factors, existing experimental evidence indicates that the complex underground storage process can be simplified by studying relative permeability as a function of saturation. Figure 2.2 shows a typical curve of gas–water two-phase relative permeability as a function of water saturation.

The relative permeability of water $Q$ is higher in regions with greater water saturation. During the initial stage of $CO_2$ displacement, the decrease in water saturation in dolomite occurs at a relatively slow rate, while the relative permeability of $CO_2$ remains at a low level. This is because $CO_2$ must overcome capillary pressure to enter, and it migrates into the smallest pore spaces through diffusion and dissolution, resulting in a very low relative permeability for $CO_2$ $k_{rw}$ at this stage. Under these slow dissolution conditions, the $CO_2$ content gradually increases, replacing the previously immobile water and further enhancing the mobility of $CO_2$. Therefore, as the displacement process continues, the residual water saturation and permeability continue to decrease, while the permeability and saturation of $CO_2$ increase.

Empirical formula for the relative permeability of homogeneous and heterogeneous porous media: Firstly, a parameter related to saturation is defined as follows:

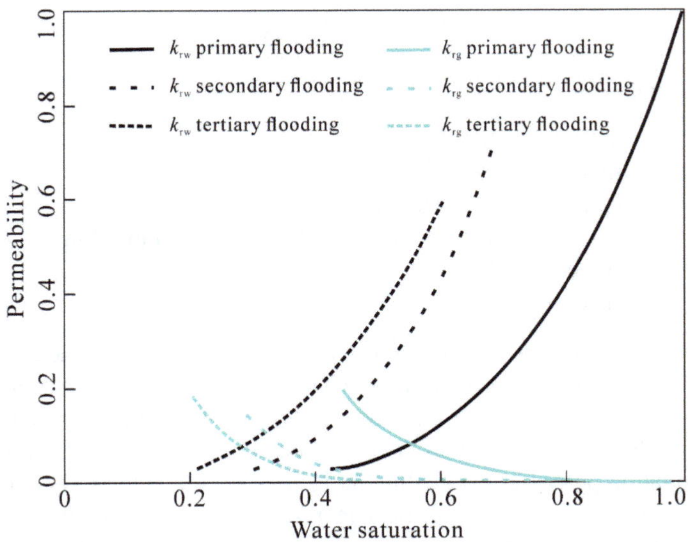


**Fig. 2.2** Relative permeability as a function of water saturation

$$\eta_w^* = \frac{S_w - S_{w\tau}}{1 - S_{w\tau}} \tag{2.8}$$

Form

$\eta_w^*$ a parameter related to saturation, which varies over time;
$S_{w\tau}$ residual water saturation (%);
$S_w$ water saturation (%)

$$\eta_w^* = (P_b/P_c)^\lambda \tag{2.9}$$

in the form

$P_c$ capillary pressure (MPa);
$P_b$ initial pressure (MPa). In porous media, $P_b$ generally refers to the pressure at which $CO_2$ first enters the porous medium and begins to displace the wetting phase (such as water);
$\lambda$ the quantitative distribution of pore sizes in a porous medium, which in theory can be any value greater than zero.

Using parameter $\eta_w^*$, the relative permeability of the two phases can be calculated as follows:

$$k_{rw} = \left(\eta_w^*\right)^{(2+3\lambda)/\lambda} \tag{2.10}$$

$$k_{rg} = \left(1 - \eta_w^*\right)^2 \left(1 - \eta_w^*\right)^{(2+\lambda)/\lambda} \tag{2.11}$$

in the form

$k_{rw}$ relative permeability to water phase;
$k_{rg}$ gas-phase relative permeability.

(2) Impact of Fractures

The storage of $CO_2$ in deep geological structures involves complex physical and chemical processes, such as multiphase and multicomponent flow in fractured reservoirs, phase transitions, $CO_2$ dissolution, mineral precipitation, and other chemical reactions. The complexity of $CO_2$ storage is further exacerbated by the presence of fractures and faults, which introduce geological heterogeneity. These fractures are present in nearly all natural reservoirs and play a key role in hydrocarbon migration and trapping. Due to their significantly higher permeability compared to the matrix rocks of the reservoir, fractures are therefore critically important in predicting the evolution and distribution of $CO_2$ gas. Fractures can create preferential flow pathways, potentially increasing the risk of $CO_2$ leakage within geological formations, which in turn may lead to significant environmental concerns. For example, after $CO_2$ is injected into the target reservoir, it may rise under buoyancy and accumulate at the base of the overlying caprock. As $CO_2$ pressure continues to increase, preferential flow may occur through faults and fractures that may exist within the caprock. In principle, geologic structures with dense faulting are not suitable for the safe storage of $CO_2$; however, fractures and faults are present in many $CO_2$ storage reservoirs at various sites. Additionally, at certain locations, hydraulic fracturing can be used to stimulate fracture growth, thereby increasing the storage capacity of $CO_2$ in saline aquifers.

The physical characteristics of fractures have a significant impact on the migration and distribution of $CO_2$. Therefore, this section discusses the influence of key fracture parameters—such as fracture length, aperture, orientation, as well as the effects of complex fracture networks—on $CO_2$ migration.

1) Fracture length refers to the length or extent of a rock fracture in a specific direction. Changes in fracture length affect the length of high-permeability flow pathways; that is, the longer the fracture, the faster the migration rate of $CO_2$. In the initial stage of $CO_2$ injection, the low-viscosity $CO_2$ displaces the original formation brine, resulting in a rapid increase in reservoir pressure. As $CO_2$ injection continues, there is sufficient space for $CO_2$ to flow, and the reservoir pressure gradually decreases. The longer the fracture, the faster the $CO_2$ migration rate, leading to less $CO_2$ accumulation near the injection point and, consequently, lower reservoir pressure. Therefore, the presence of fractures reduces pressure build-up near the $CO_2$ wellhead, thereby enhancing the safety of $CO_2$ storage. Within the same reservoir, the longer the fracture, the more significant the alleviation of reservoir pressure.
2) Fracture aperture refers to the distance between the two walls of a fracture and is an important parameter characterising fracture properties. An increase in fracture aperture leads to a higher fracture permeability, thereby accelerating the migration rate of $CO_2$. As the migration rate of $CO_2$ increases, the accumulation of $CO_2$ near the injection point is reduced. With continued $CO_2$ injection, the brine surrounding the injection well is gradually displaced; once the brine around the injection well has been fully displaced, there is sufficient space for $CO_2$ migration, making subsequent $CO_2$ injection relatively easier. In contrast, fractures with smaller apertures have a weaker capacity for $CO_2$ migration, resulting in a large proportion of the $CO_2$ plume migrating only within the matrix, and the pressure build-up near the injection well is difficult to alleviate. Therefore, the larger the fracture aperture, the more significant the relief of reservoir pressure.
3) Fracture orientation refers to the direction in which rock fractures extend at the surface or underground. Variations in fracture orientation influence both the migration direction of $CO_2$ plumes and the pressure distribution within the reservoir. When fracture orientations are relatively consistent and fracture density is high, $CO_2$ tends to migrate along the direction of the fractures, forming more concentrated migration pathways. This mode of migration facilitates the rapid dispersion and effective sealing of $CO_2$. However, when fracture orientations are irregular or fracture density is low, the migration direction of $CO_2$ becomes more complex and difficult to predict, potentially resulting in uneven distribution

of $CO_2$ within the underground storage formation and reducing the reservoir's capacity to securely contain $CO_2$. Notably, when the injection direction forms a 45° angle with the fracture orientation, the $CO_2$ migration rate is fastest; at 0° and 90°, the rate is moderate, and at 135°, it is slowest.

4) Complex fracture networks have a significant impact on the migration pathways, rates, and ultimate storage locations of $CO_2$. When a complex fracture network exists within the reservoir, it greatly increases the heterogeneity of the reservoir, forming a network of high-permeability channels that accelerate the migration rate of $CO_2$ and alleviate reservoir pressure. However, the presence of a complex fracture network also increases the instability of the reservoir and enhances the uncertainty of $CO_2$ storage, which may lead to issues such as $CO_2$ leakage.

In addition, some potential $CO_2$ storage reservoirs have already been found to possess insufficient porosity and permeability, making it impossible to support economically viable commercial-scale injection. On this basis, the application of hydraulic fracturing technology can enhance both the injectivity and storage capacity of $CO_2$ reservoirs. Modelling studies indicate that fracturing can moderately increase reservoir capacity (by 10–35%). The simulation results of this study confirm that, within the range of horizontal well length, number of fractures, fracture geometry, and fracture properties, the use of fracturing technology consistently improves injectivity. Compared with unfractured vertical wells, oil and gas production capacity is increased by 13–71%.

### *2.1.3 Geological Conditions for Carbon Dioxide Storage*

While $CO_2$ is used to displace oil and gas, it is simultaneously sequestered underground, primarily retained within the pore spaces of geological formations. The selection of storage sites should fully consider factors such as the stability of geological structures, the properties and storage capacity of the reservoir, and the integrity of the caprock. As shown in Fig. 2.3, favourable closure conditions are a prerequisite for the safe and effective storage of $CO_2$. Stable geological structures ensure the security and integrity of the $CO_2$ storage space, preventing breaches; the properties of the reservoir directly affect the available storage capacity and the efficiency of $CO_2$ storage; and a stable caprock ensures that, following $CO_2$ injection, upward migration and leakage are prevented. At present, the application of $CO_2$ storage remains at the industrial pilot stage, and there is not yet a specific standard for the evaluation of storage sites. Based on the conditions required for oil and gas closures, the selection of $CO_2$ storage sites mainly focuses on the following aspects:

(1) Simple Structure or Area Distant from Complex Structural Zones

The presence of faults and similar structures provides pathways for $CO_2$ diffusion, leading to $CO_2$ leakage. At the same time, these structures disrupt the continuity of

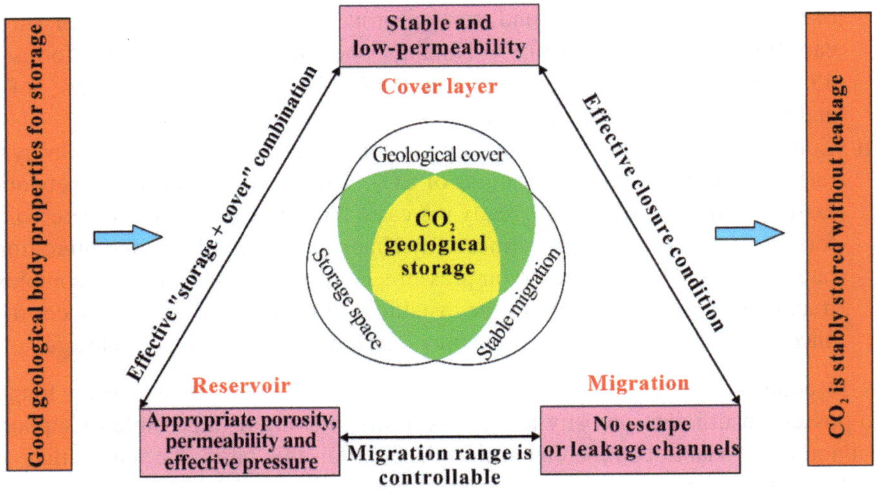


**Fig. 2.3** Geological conditions for $CO_2$ storage

the reservoir, which is unfavourable for the lateral migration of $CO_2$, resulting in a reduction in displacement efficiency.

(2) Favourable Enclosure Conditions

The reservoir exhibits lateral stability and is not vertically connected with other strata, ensuring the presence of an effective caprock above the storage formation, such as a mudstone or shale layer.

(3) Appropriate Depth

When the reservoir is too shallow, the lithostatic pressure is relatively low, and $CO_2$ exists in a gaseous state. At this time, the storage capacity of $CO_2$ is limited and it is closer to the surface, resulting in greater safety risks. Conversely, when the reservoir is too deep, the lithostatic pressure is relatively high, the reservoir permeability is low, and the migration rate of $CO_2$ is restricted.

(4) Moderate Permeability

An appropriate permeability ensures a suitable migration rate for $CO_2$, and is also a key parameter for evaluating the enhancement of recovery rates through $CO_2$ flooding.

(5) No Engineering Disturbance in the Vicinity

For a certain period or over an extended duration near the $CO_2$ storage site, no development or extraction activities should be conducted, in order to protect the integrity of the reservoir caprock and maintain the stability of the stress field.

## 2.2 Fluid Replacement Theory

Petrophysics serves as the bridge connecting the composition of rocks (petrophysical parameters) with elastic parameters. In $CO_2$ storage, by analysing the mineral composition, pore structure, pore fluids, and pressure of the target reservoir rock, it is possible to calculate parameters such as the reservoir's P-wave velocity, S-wave velocity, and density. During different stages of $CO_2$ injection, pore fluids undergo substitution, resulting in changes in P-wave and S-wave velocities as well as density. These variations form the basis for monitoring $CO_2$ migration using time-lapse seismic methods. Therefore, this section introduces the fluid substitution theory based on the Gassmann equation, outlines the fundamental theory of the Gassmann equation, and describes the calculation methods for rock frame and fluid bulk modulus models, thereby providing a theoretical foundation for simulating changes in reservoir elastic parameters following $CO_2$ saturation.

### *2.2.1 Gassmann Equation*

The Gassmann equation establishes the relationship between the pore fluid in a reservoir rock and the elastic parameters of the rock by calculating the parameters of saturated rocks. Based on the Gassmann equation, various physical parameters of reservoir rocks can be obtained, such as the rock frame modulus, fluid bulk modulus, P-wave and S-wave velocities, density, porosity, and Poisson's ratio. The Gassmann equation provides a theoretical foundation for calculating the elastic parameters of reservoirs under different fluid saturations and pressure conditions. Therefore, it is possible to simulate and obtain the P-wave and S-wave velocities at different stages of $CO_2$ displacement.

The Gassmann equation can be expressed as follows:

$$K_{sat} = K_{dry} + \frac{\left(1 - \frac{K_{dry}}{K_m}\right)^2}{\frac{\phi}{K_{fl}} + \frac{1-\phi}{K_m} - \frac{K_{dry}}{K_m^2}} \tag{2.12}$$

$$\mu_{sat} = \mu_{dry} \tag{2.13}$$

In the formula, $K_{sat}$, $K_{dry}$, $K_m$, and $K_{fl}$ represent the bulk moduli of saturated sandstone, dry sandstone, sandstone matrix, and pore fluid, respectively; $\mu_{sat}$ and $\mu_{dry}$ denote the shear moduli of saturated sandstone and dry sandstone, respectively; $\phi$ is the porosity.

The velocities of compressional and shear waves are related to the bulk modulus and shear modulus of rocks, as follows:

$$\rho v_p^2 = K_{sat} + \frac{4}{3}\mu_{dry} \tag{2.14}$$

$$\rho v_s^2 = \mu_{sat} = \mu_{dry} \tag{2.15}$$

In the equation, $v_p$, $v_s$, and $\rho$ represent the saturated sandstone P-wave velocity, S-wave velocity, and density, respectively. By applying Eqs. (2.14) and (2.15), the corresponding P-wave velocity, S-wave velocity, and density under different saturation conditions can be determined.

The Gassmann equation provides a convenient means for calculating the elastic parameters of reservoirs; however, its application requires the fulfilment of several conditions: (1) at the macroscopic scale, the rock matrix and framework must be homogeneous; (2) all pores within the rock must be fully interconnected; (3) the pore space must be saturated with fluid and free from shear resistance; (4) the rock-fluid system must be closed, with no fluid exchange with the external environment; and (5) no physical or chemical reactions occur between the fluid and the rock framework.

From Eq. (2.12), it can be seen that the Gassmann equation incorporates information on both the rock matrix and the pore fluid. During $CO_2$ saturation and storage, $CO_2$ is injected into the reservoir in a supercritical state, causing changes in the pore pressure of the reservoir, which in turn leads to variations in the bulk modulus of the dry rock matrix $v_p$. After $CO_2$ injection, the composition of the pore fluid changes, resulting in a variation in the fluid bulk modulus $v_s$. Furthermore, $CO_2$ may react chemically with the reservoir cement or the rock matrix, dissolving certain mineral components and thereby altering the reservoir porosity $\rho$. Therefore, the process of $CO_2$ saturation exceeds the basic assumptions of Gassmann's theory. In such cases, it is necessary to re-evaluate the changes in the dry rock bulk modulus $K_{dry}$ and porosity $K_{fl}$ in the calculations.

## 2.2.2 *Calculation of Rock Matrix Modulus*

In general, the rock matrix is composed of various minerals. In the calculation of the modulus of the rock matrix $K_{dry}$, the Voigt and Reuss models analyse the mixing rules of different minerals, and use the moduli of the mineral components to estimate the modulus of the rock matrix. The formulas are as follows:

Voigt Equation:

$$M_V = \sum_{i=1}^{N} f_i \times M_i \tag{2.16}$$

Reuss Formula:

$$\frac{1}{M_R} = \sum_{i=1}^{N} \frac{f_i}{M_i} \tag{2.17}$$

In the equations, $M_V$ represents the upper bound of the sandstone framework modulus calculated using the Voigt method, $M_R$ denotes the lower bound of the sandstone framework modulus calculated using the Reuss method, $i$ indicates the mineral component, $N$ represents the total number of minerals, and $f_i$ denotes the percentage content of the $i$th mineral. Subsequently, Hill used Eqs. (2.16) and (2.17) to calculate the arithmetic mean of the sandstone framework modulus, achieving a better result, which is referred to as the Voigt–Reuss–Hill (V-R-H) model. Therefore, the arithmetic mean of the elastic modulus of the sandstone framework can be defined as:

$$M_{VRH} = \frac{M_V + M_R}{2} \tag{2.18}$$

The V-R-H method is currently a commonly used approach for calculating the framework modulus of rocks with multiple mineral components.

### 2.2.3 Calculation of Fluid Bulk Modulus

When a single-phase fluid is present in the reservoir, the bulk modulus of the fluid is typically calculated using empirical formulas derived from the laboratory experiments of Batale and Wang. Based on the parameters of oil, gas, and water, the elastic modulus of each phase fluid under different reservoir conditions can be determined. When multiphase fluids are present in the reservoir, the bulk modulus of the fluid is commonly calculated using the Wood model, Patchy model, Domenico model, and the empirical formula of Brie, depending on the specific geological characteristics and requirements of the task. Among these, the Wood model assumes that the minerals and rocks are elastic and that the fluid mixture is homogeneous, making it suitable for calculating the bulk modulus of multiphase fluids in high-porosity, high-permeability reservoirs. The model can be defined by the following formula:

$$\frac{1}{K_{fl}} = \frac{S_w}{K_w} + \frac{S_g}{K_g} + \frac{S_o}{K_o} \tag{2.19}$$

$$\rho_{fl} = S_w\rho_w + S_g\rho_g + S_o\rho_o \tag{2.20}$$

In the equation, $K_w$, $K_o$ and $K_g$ represent the volumetric moduli of water, oil, and gas ($CO_2$), respectively; $S_w$, $S_o$ and $S_g$ denote the saturations of water, oil, and gas ($CO_2$), respectively; $\rho_{fl}$, $\rho_w$, $\rho_g$ and $\rho_o$ represent the densities of the mixed fluid, water, oil, and gas ($CO_2$), respectively.

In the Domenico model, it is assumed that the pores are not interconnected for different fluids, and the fluid volume is represented as a weighted function of the multiphase fluid volumes:

$$K_{fl} = K_w S_w + K_o S_o + K_g S_g \tag{2.21}$$

The patchy model is proposed based on the condition of patchy saturation, and through simplification, it can yield an expression consistent with the Domenico model. Based on extensive experimental data analysis, Brie et al. established an empirical formula for calculating the fluid volume fraction:

$$K_{fl} = \left(K_w - K_g\right) S_w^e + K_g \tag{2.22}$$

In the equation, $e$ represents the empirical coefficient, which is determined based on statistical analysis of empirical data.

In practical applications, the calculation of fluid volumetric modulus requires a comprehensive determination based on actual geological conditions and rock physical experimental data.

### 2.2.4 *Analysis of Reservoir Elastic Parameter Changes After Fluid Replacement*

Based on the fluid substitution theory introduced earlier, this section uses actual well log curves from a certain oilfield as an example to calculate the variations in reservoir elastic parameters during different periods of $CO_2$ saturation. The well in question is planned to serve as a $CO_2$ injection well; however, the $CO_2$ flooding test has not yet been implemented. By simulating the changes in P-wave and S-wave velocities and density under different $CO_2$ saturation conditions, theoretical parameters can be provided for modelling $CO_2$ migration monitoring. These results can also be used to analyse the feasibility of monitoring migration using time-lapse seismic methods.

As shown in Fig. 2.4, the well logging curves for the $CO_2$ injection well include compressional wave velocity, shear wave velocity, density, porosity, and water saturation curves. The logging depth ranges from 480 to 1265 m. The red box indicates the planned $CO_2$ injection interval, with a depth of approximately 1145–1165 m.

Based on the above actual well log curves, the elastic parameters under different $CO_2$ saturation conditions can be calculated using Gassmann's fluid substitution theory. The following describes the implementation process of fluid substitution:

(1) Based on the Wood model, fluid density and fluid bulk modulus are calculated using Eqs. (2.19) and (2.20).
(2) Calculation of the bulk modulus of dry rock $e$. Given the known P-wave and S-wave velocities and density of saturated rock, at this point:

$$\mu_{sat} = \rho_b V_s^2 \tag{2.23}$$

$$M_{sat} = \rho_b V_p^2 \tag{2.24}$$

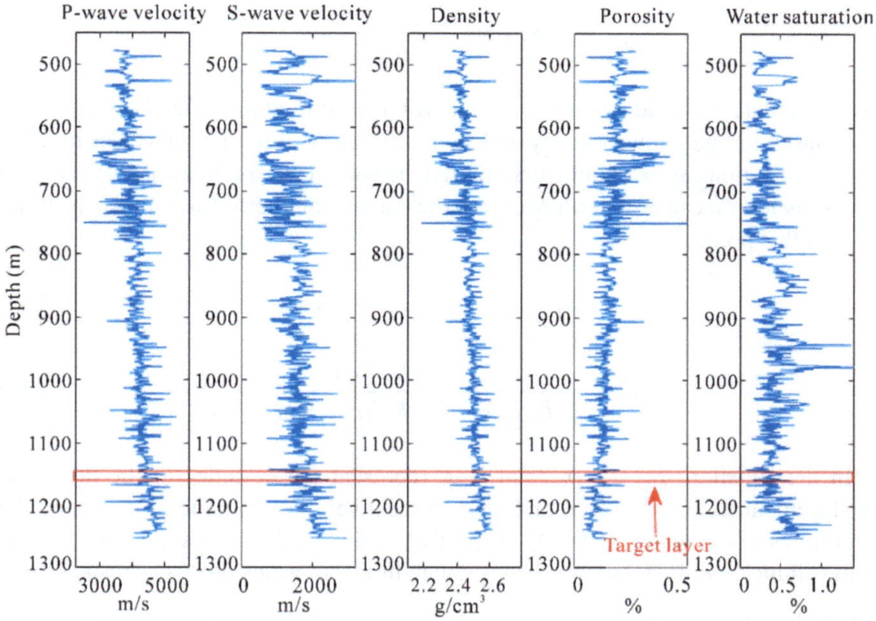


**Fig. 2.4** Well-logging curves of the $CO_2$ injection well

In the equation, $\rho_b$ denotes the density of saturated sandstone, $V_p$ and $V_s$ represent the P-wave and S-wave velocities of saturated sandstone, respectively, and $\mu_{sat}$ and $M_{sat}$ are the shear modulus and elastic modulus of saturated sandstone, respectively, where $\mu_{sat} = \mu_{dry}$. Accordingly, the effective bulk modulus of saturated sandstone, $K_{sat}$, can be obtained as follows:

$$K_{sat} = M_{sat} - \frac{4}{3}\mu_{sat} \tag{2.25}$$

To calculate the bulk modulus of dry rock $K_{dry}$, the variables $a$ and $b$ are introduced here:

$$a = \phi\frac{K_m}{K_{fl}} + 1 - \phi \tag{2.26}$$

$$b = \phi\frac{K_m}{K_{fl}} + \frac{K_{sat}}{K_m} - 1 - \phi \tag{2.27}$$

In the equation, $K_m$ and $K_{fl}$ represent the bulk modulus of the matrix and the fluid, respectively; $\phi$ denotes porosity, which can be obtained from well logging curves. At this point, $K_{dry}$ can be expressed as:

$$K_{dry} = \frac{aK_{sat} - K_m}{b} \tag{2.28}$$

(3) Calculate the fluid density and fluid volumetric modulus after $CO_2$ injection. Based on the known density and saturation of the fluid, the fluid density after substitution can be obtained. Similarly, the volumetric modulus of the fluid after substitution can be determined from the known volumetric modulus of the fluid. At this point:

$$\rho'_{fl} = S_o\rho_o + S_w\rho_w + S_{CO_2}\rho_{CO_2} \tag{2.29}$$

$$K'_{fl} = \frac{K_o K_w K_{CO_2}}{S_o K_w K_{CO_2} + S_w K_o K_{CO_2} + S_{CO_2} K_o K_w} \tag{2.30}$$

In the equation, $\rho'_{fl}$, $\rho_o$, $\rho_w$, and $\rho_{CO_2}$ represent the densities of the post-$CO_2$ displacement fluid, oil, water, and $CO_2$, respectively; $S_o$, $S_w$, and $S_{CO_2}$ denote the saturations of oil, water, and $CO_2$, respectively; $K'_{fl}$, $K_o$, $K_w$, and $K_{CO_2}$ represent the fluid volume moduli of the post-displacement fluid, oil, water, and $CO_2$, respectively.

(4) Calculate the density of the reservoir rock and the velocities of the P-wave and S-wave after substitution, based on the well log saturation curves and the density of the substituted fluid $\rho'_{fl}$. The density of the $CO_2$-saturated reservoir rock after substitution can be expressed as:

$$\rho'_b = (1 - \phi)\rho_m + \phi\rho'_{fl} \tag{2.31}$$

In the equation, $\rho'_b$ represents the density of the replaced saturated tuff, $\rho_m$ denotes the average density of the tuff matrix, and $\rho'_{fl}$ refers to Eq. (2.29).

The saturated dolomite volume after $CO_2$ replacement $K'_{sat}$ can be expressed as:

$$K'_{sat} = K_{dry} + \frac{\left(1 - \frac{K_{dry}}{K_m}\right)^2}{\frac{\phi}{K'_{fl}} + \frac{1-\phi}{K_m} - \frac{K_{dry}}{K_m^2}} \tag{2.32}$$

According to Eq. (2.25), the saturated dolomite elastic modulus after $CO_2$ replacement $M'_{sat}$ can be expressed as:

$$M'_{sat} = K'_{sat} + \frac{4}{3}\mu_{sat} \tag{2.33}$$

From Eqs. (2.23) and (2.24), the longitudinal and shear wave velocities after $CO_2$ replacement can be expressed as follows:

$$V'_p = \sqrt{\frac{M'_{sat}}{\rho'_b}} \tag{2.34}$$

$$V_s' = \sqrt{\frac{\mu_{sat}}{\rho_b'}} \tag{2.35}$$

Based on the above fluid substitution process, using the fluid parameters shown in Table 2.1, the post-substitution well log curves can be obtained. In this simulation, the water saturation is derived from the well log curves and remains unchanged before and after substitution, with $CO_2$ replacing 70% of the original oil content. To highlight the changes in elastic parameters of the target layer following $CO_2$ injection, the difference between the pre- and post-substitution curves is taken to obtain the variation curve, as shown in Fig. 2.5. It can be observed that when $CO_2$ saturation reaches 70%, the maximum change in P-wave velocity is nearly − 200 m/s, while the minimum change is approximately − 100 m/s. This provides a theoretical basis for using time-lapse seismic data to monitor the migration of $CO_2$.

**Table 2.1** Fluid substitution parameters for $CO_2$ displacement

| Fluid type | Density (g/cc) | Velocity (m/s) | Bulk modulus (MPa) |
|---|---|---|---|
| Oil | 0.75 | 1154.70 | 1000 |
| Water | 1.09 | 1477.66 | 2380 |
| $CO_2$ | 0.13 | 561.59 | 41 |

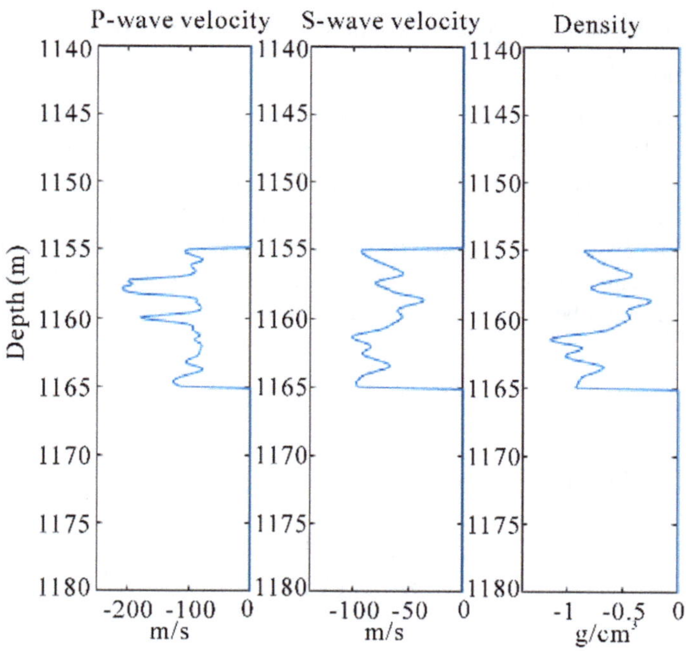


**Fig. 2.5** Variation in elastic parameters after $CO_2$ substitution

## 2.3 Multi-phase Multi-field Coupling Theory

In processes such as $CO_2$-enhanced oil and gas recovery, coalbed methane extraction, and aquifer storage, the injected fluids interact with the formation fluids, resulting in flow dispersion. Simultaneously, the flow regime of the fluids exists within a complex multiporous medium, where temperature fields, mechanical fields, and other multiple fields are intricately coupled. This section provides a comprehensive analysis of the effects of flow dispersion and the čoupling of flow and solid mechanics, establishing control equations for multiphase, multicomponent seepage fields, stress–strain fields, and temperature fields, as well as fully coupled multi-field control equations that account for flow dispersion. In addition to considering the conventional coupling between fluid and rock matrix stress–strain during the coupling process, the coupling of the temperature field and the unique dispersion effects of multicomponent fluids are also taken into account.

### 2.3.1 *Seepage Field Equation*

In practice, geological storage of $CO_2$ involves multiphase flow in porous media, where the effects of gravity during the flow process cannot be neglected. In the context of $CO_2$-enhanced oil recovery, due to the complex composition of crude oil, the actual multiphase flow process is a multicomponent, multiphase flow.

(1) Equation of Motion

Assume that the fluid contains N oil components and N gas components, resulting in a total of 2N + 1 displacement variables, namely the displacement vector of the solid particles $u_s$ and the displacement vectors of each component in the oil and gas phases, $u_{oi}$ and $u_{gi}$ (with units of m). Correspondingly, there are 2N + 1 velocities $v_s$, $v_{oi}$, and $v_{gi}$ (with units of m/s). The subscripts $s$, $o$, and $g$ correspond to the solid skeleton, oil, and gas, respectively. Let the relative velocities of oil and gas with respect to the solid be $v_{ori}$ and $v_{gri}$, respectively, i.e.

$$\left.\begin{array}{l} v_{oir} = v_{oi} - v_s \\ v_{gir} = v_{gi} - v_s \end{array}\right\} (i = 1, 2, 3, \ldots, N) \tag{2.36}$$

Therefore, this equation can be written as:

$$\left.\begin{array}{l} v_{oir} = -\frac{KK_{roi}}{\mu_{oi}\phi}(\nabla p + \rho_{oi} g) \\ v_{gir} = -\frac{KK_{rgi}}{\mu_{gi}\phi}(\nabla p + \rho_{gi} g) \end{array}\right\} (i = 1, 2, 3, \ldots, N) \tag{2.37}$$

In formula:

$\rho_{oi}$, $\rho_{gi}$ represent the densities of the oil and gas components, respectively, in kg/m$^3$;

$K$ absolute permeability, $\mu m^2$;
$K_{roi}, K_{rgi}$ represent the relative permeabilities of oil and gas phase components, respectively;
$\mu_{oi}, \mu_{gi}$ represent the viscosities of the oil and gas components, respectively, in mPa s;
$\phi$ porosity, unitless;
$\nabla p$ pressure gradient, MPa/m;
$g$ gravitational acceleration, generally taken as 9.8 m/s$^2$.

(2) Continuity Equation

Disregarding the effects of sources and sinks, the continuity equation for oil and gas is as follows:

$$\left.\begin{array}{l}\frac{\partial(\rho_{oi}S_{oi}\phi)}{\partial t}+\nabla(\rho_{oi}\phi v_{oi})=0\\ \frac{\partial(\rho_{gi}S_{gi}\phi)}{\partial t}+\nabla(\rho_{gi}\phi v_{gi})=0\end{array}\right\}(i=1,2,3,\ldots,N) \tag{2.38}$$

In the formula:

$S_{oi}$ saturation of $i$ component in the oil phase, unitless;
$S_{gi}$ saturation of the $i$ component in the gas phase, dimensionless.

The continuity equation for solids is

$$\frac{\partial[(1-\phi)\rho_s]}{\partial t}+\nabla[(1-\phi)\rho_s v_s]=0 \tag{2.39}$$

(3) Fundamental Differential Equations of Seepage Fields

Based on the assumption of small deformations, $\varphi$, $\rho_{oi}$, $\rho_{gi}$ and $\rho_s$ can be expressed as:

$$\varphi=\varphi_a\left[1+c_\varphi(p-p_a)\right]=\varphi_a\left(1+c_\varphi\Delta p\right) \tag{2.40}$$

$$\rho_{oi}=\rho_{oai}\left[1+c_{oti}(p-p_a)\right]=\rho_{oai}(1+c_{oti}\Delta p) \tag{2.41}$$

$$\rho_{gi}=\rho_{gai}\left[1+c_{gti}(p-p_a)\right]=\rho_{gai}\left(1+c_{gti}\Delta p\right) \tag{2.42}$$

$$\rho_s=\rho_{sa}\left[1+\frac{p-p_a}{K_m}-\frac{tr\left(\sigma'-\sigma_a'\right)}{(1-\phi)3K_m}\right] \tag{2.43}$$

In the formula:

$\varphi_a$ initial porosity of the rock, unitless;
$p$ confining pressure in the stratum, MPa;
$p_a$ initial lithostatic pressure, MPa;

$\rho_{oai}$ oil phase $i$ component initial density, kg/m$^3$;
$\rho_{gai}$ initial density of gas-phase $i$ component, kg/m$^3$;
$c_\varphi$ pore compressibility coefficient, MPa$^{-1}$;
$c_{oti}$ oil phase $i$ component density compressibility coefficient, MPa$^{-1}$;
$c_{gti}$ compressibility coefficient of the $i$ component in the gas phase, MPa$^{-1}$;
$c_{gti}$ initial density of rock, kg/m$^3$;
$K_m$ density coefficient of rock, kg/m$^3$;
$tr$ vector function, dimensionless;
$\sigma_a'$ initial stress, MPa;
$\sigma_a'$ effective stress, MPa.

The second and third terms on the right-hand side of Eq. (2.43) represent, respectively, the variations in solid density caused by fluid pressure and effective stress. The effective stress $\sigma_a'$ is given by the overall continuity equation.

Substituting Eq. (2.37) and $\rho_s$ with respect to t into the continuity equation, and rearranging, the differential equation for seepage flow is obtained as follows:

$$\rho_{oi}\phi\frac{\partial S_{oi}}{\partial t}+\rho_{gi}\phi\frac{\partial S_{gi}}{\partial t}+\left[\frac{\rho_{sai}}{K_m}\left(1-\frac{1}{K_m}-\phi\right)+S_{oi}\phi\rho_{gai}c_{gti}\right]\frac{\partial p}{\partial t}$$
$$+\left[(1-\phi)\rho_{sai}+\rho_{oi}\phi+\rho_{gi}\phi-\frac{\rho_{sai}}{K_m}\right]\frac{\partial\varepsilon_v}{\partial t}+\nabla\left[\frac{\rho_{oi}K_{oi}}{\mu_{oi}}(\nabla p+\rho_{oi}g)\right] \tag{2.44}$$
$$+\nabla\left[\frac{\rho_{gi}K_{gi}}{\mu_{gi}}(\nabla p+\rho_{gi}g)\right]=0$$

Formula:

$\varepsilon_v$ deformation of the marble framework, m.

### 2.3.2 Convective-diffusion Equation

During the $CO_2$ miscible flooding process, $CO_2$ comes into contact with crude oil either multiple times or just once, thereby forming a miscible phase. However, in the displacement process, due to the mass transfer effects between $CO_2$ and crude oil, $CO_2$ in certain regions may undergo significant dilution, even to the extent of reaching the two-phase region, thus transitioning to an immiscible displacement stage. In 2002, Johns et al. found that the reduction in recovery efficiency caused by dispersion and dilution could reach up to 15%. Therefore, it is necessary to consider the impact of dispersion and mass transfer on the effectiveness of displacement.

The advection–diffusion equation is a fundamental type of motion equation, derived by performing a material balance on an infinitesimal element within the system. For a multi-component system, the amount of a given component entering an infinitesimal element, plus the amount generated within the element by chemical reactions, minus the amount leaving the element, equals the accumulation of that component within the element. Considering that the influx and efflux of the component are caused by both diffusion and advection, and that the diffusive flux is described by Fick's law, the advection–diffusion equation can thus be expressed as follows:

$$\frac{\partial \phi C}{\partial t} + \nabla[-D(v)\nabla C + Cv] = 0 \tag{2.45}$$

In the equation: $D(v)$ represents the molecular diffusion coefficient at a given velocity. Substituting the respective parameters into the advection–diffusion equation yields:

$$C_{gi}\Delta p\frac{\partial^2\phi}{\partial p\partial t} + C_{gi}\frac{\partial \phi}{\partial t} + \phi_a C_\phi \Delta p\frac{\partial C_{gi}}{\partial t} + \nabla\left[-D_{ij}\nabla C_{gi} - \frac{C_{gi}KK_{rgi}}{\mu_{gi}\phi}(\nabla p + \rho_{gi}g\nabla z)\right] = 0 \tag{2.46}$$

In the equation, $D_{ij}$ represents the molecular diffusion coefficient of component $i$ in component $j$, $C_{gi}$ denotes the mass concentration of component $i$, and $t$ represents time. The propagation of the medium is closely related to the flow; only after obtaining the velocity distribution can the concentration distribution be derived from the advection–diffusion equation, and subsequently, the flux of medium propagation can be determined.

### 2.3.3 Fundamental Equations of Rock Deformation

The establishment of the rock deformation relationship is based on the assumption of linear elasticity. That is, the total strain is equal to the sum of the strain induced by confining pressure and the strain caused by stress. An increase in fluid pressure within the pores will also induce strain. For isotropic solid media, the strains along the three principal axes are equal and do not result in shear strain. The volumetric strain is given by $\varepsilon_p = -\frac{1}{K_m}(p - p_a)$, while the uniaxial strain $\varepsilon_p$ is

$$\varepsilon_p = -\frac{1}{3K_m}(p - p_a) \tag{2.47}$$

The total strain of the solid framework is

$$\varepsilon = -\frac{1}{2G}\sigma' - \varepsilon_p\frac{\gamma}{2G(1+\gamma)}tr(\sigma') \tag{2.48}$$

In the equation: $G$ denotes the shear modulus, $\varepsilon_p$ represents the strain induced by compression, and $\gamma$ denotes the shear strain.

The two terms on the right-hand side of Eq. (2.48) represent the strain induced by stress. From Eq. (2.48), it is straightforward to derive the expression for stress in terms of strain:

$$\sigma' = 2G\varepsilon + \frac{\gamma}{1+\gamma} tr(\varepsilon) - 3\varepsilon_p \tag{2.49}$$

Here, the effective stress $\sigma$ consists of two components: the macroscopic stress $\sigma$ and the internal pressure $P$, i.e., $\sigma' = \sigma + P$. Therefore, the fundamental constitutive equation for the fluid–solid composite can be written as:

$$\sigma = 2G\varepsilon + \frac{\gamma}{1+\gamma} tr(\varepsilon) - \alpha P \tag{2.50}$$

Among them,

$$\alpha = 1 - \frac{1}{K_m}$$

Using the Darcy equation $2\varepsilon_{ij} = 2G(\mu_{i,j}) = G(\mu_{i,j} + \mu_{j,i})$, Eq. (2.50) can be expressed in index notation as follows:

$$\sigma_{ij} = G(\mu_{i,j} + \mu_{j,i}) + \lambda\mu_{k,k} - \alpha p\delta_{ij}$$

In the formula: $\sigma_{ij}$ denotes the total stress.

Taking the derivative of Eq. (2.50) with respect to $x_j$ yields $(i = x, y, z)$. Based on the equilibrium state of the unit body under force, the equilibrium equation is obtained as follows:

$$\sigma_{ij,j} + f_i = 0 \tag{2.51}$$

In the equation: $f_i = 0$ represents the volumetric force. According to the effective stress principle, the total stress can be expressed in terms of effective stress as follows:

$$G\nabla^2\mu_i + (G+\lambda)\varepsilon_{vi} - ap_i\delta_{ij} + f_i = 0 \tag{2.52}$$

Taking the time $t$ derivative again, we obtain

$$f_z = -[\phi s_o\rho_o + \phi s_g\rho_g + (1-\phi)\rho_s] \tag{2.53}$$

$$f_x = f_y = 0 \tag{2.54}$$

By substituting the basic equations for density and porosity, the final set of fundamental equations governing fluid–solid coupling is obtained as follows:

$$G\frac{\partial}{\partial t}(\nabla^2\mu_x) + (G+\lambda)\frac{\partial^2\varepsilon_v}{\partial t\partial x} - \alpha\frac{\partial^2 p}{\partial t\partial x} = 0 \tag{2.55}$$

$$G\frac{\partial}{\partial t}(\nabla^2\mu_y) + (G+\lambda)\frac{\partial^2\varepsilon_v}{\partial t\partial y} - \alpha\frac{\partial^2 p}{\partial t\partial y} = 0 \tag{2.56}$$

$$G\frac{\partial}{\partial t}(\nabla^2\mu_z) + (G+\lambda)\frac{\partial^2\varepsilon_v}{\partial t\partial z} - \alpha\frac{\partial^2 p}{\partial t\partial z} - \rho_{sa}(1-\alpha)\frac{\partial\varepsilon_v}{\partial t}$$
$$-\left[\rho_{oai}\phi c_{oi} + \rho_{gai}\phi c_{gi} + \rho_{sa}\left(\frac{\alpha-\phi}{K_m} - \phi_a c_\phi\right)\right]g\frac{\partial p}{\partial t} = 0 \quad (i = 1, 2, 3, \ldots, N) \tag{2.57}$$

### 2.3.4 Temperature Field Equation

According to the first law of thermodynamics, within a unit volume over a unit time, the sum of the energy transferred into the system from the external environment and the heat generated internally is equal to the sum of the increase in the internal energy of the substance and the work done on the surroundings. In the case of non-isothermal conditions, neglecting viscous dissipation, the energy equation for a single-phase fluid is:

$$\phi(\rho_l c_{lV})\frac{\partial T_l}{\partial t} + (\rho_l c_{lV})(\phi V_l \cdot \nabla T_l) = -\nabla\cdot\phi J_l - \left(\frac{\partial P}{\partial T_l}\right)\phi T_l \nabla\cdot V_l + \phi q_{lk} \tag{2.58}$$

Disregarding the elastic properties of the rock skeleton, the energy equation for the rock skeleton medium is as follows:

$$(1-\phi)\rho_s c_{sV}\left(\frac{\partial T_s}{\partial t} + V_s\nabla T_s\right) = -\nabla\cdot(1-\phi)J_s - (1-\phi)\beta_{Tm}T_s\frac{\partial\varepsilon_V}{\partial t} + (1-\phi)q_{sk} \tag{2.59}$$

$\phi(\rho_l c_{lV})$ represents the rate of change of internal energy per unit volume per unit time, i.e., the internal energy accumulation term; $\phi V_l \cdot \nabla T$ denotes the difference between the energy flowing into and out of the control volume, i.e., the advection term; $\nabla\cdot\phi J_l$ is the heat conduction term; $\nabla\cdot\phi J_l$ indicates the internal energy generated per unit time, i.e., the heat source intensity. In the absence of sources and sinks, this term is zero. $\beta_{Tm}T_s\frac{\partial\varepsilon_V}{\partial t}$ is the thermoelastic energy term; $T$ denotes temperature; $c_{lV}$ and $c_{sV}$ represent the specific heat capacities of the fluid and solid, respectively; $J_l$ and $J_s$ denote the thermal densities of the fluid and solid, respectively. According to Fourier's law, $\left.\begin{array}{l} J_l = -K_l\nabla T_l \\ J_s = -K_s\nabla T_s \end{array}\right\}$, $K_l$ and $K_s$ are the thermal conductivities of the fluid and solid, respectively.

By adding the fluid energy Eq. (2.56) to the solid matrix energy Eq. (2.57), the combined energy equation for single-phase fluid heat-flow-deformation coupling is obtained as follows:

$$
\begin{aligned}
&(\rho_c c_V)_m \frac{\partial T}{\partial t} + [\phi \rho_l c_{lV} V_l + (1-\phi)\rho_s c_{sV} V_s]_l \cdot \nabla T \\
&= \nabla \cdot K_t \nabla T + \phi T \frac{\beta_l}{c_l} kP - (1-\phi)\beta_{Tm} \frac{\partial \varepsilon_V}{\partial t} + q_m
\end{aligned}
\tag{2.60}
$$

In the formula:

$$(\rho_c c_V)_m = \phi \rho_l c_{lV} + (1-\phi)\rho_s c_{sV}$$

$$K_t = \phi K_l + (1-\phi) K_s$$

$$q_m = \phi q_{lk} + (1-\phi) q_{sk}$$

$$(1-\phi)\beta_{Tm} \frac{\partial \varepsilon_V}{\partial t} = \frac{1}{2}(1-\phi)\beta_{Tm} T \frac{\partial}{\partial t}\left(\mu_{i,j} + u_{j,i}\right)\delta_{ij}$$

Disregarding the effects of sources and sinks, the final energy equation obtained is:

$$
\begin{aligned}
&(\rho_c c_V)_m \frac{\partial T}{\partial t} + [\phi \rho_l c_{lV} V_l + (1-\phi)\rho_s c_{sV} V_s]_l \cdot \nabla T \\
&= \nabla \cdot K_t \nabla T + \phi T \frac{\beta_l}{c_l} \nabla V_l - (1-\phi)\beta_{Tm} \frac{\partial \varepsilon_V}{\partial t}
\end{aligned}
\tag{2.61}
$$

The governing partial differential equations for pore fluid flow, rock matrix deformation, temperature field, and dispersion effects on flow have been presented above. These constitute the fundamental set of control equations for coupled $CO_2$ multiphase field models. However, it is essential to specify the necessary initial and boundary conditions in order to solve these equations.

### 2.3.5 *Cross-coupling Relation*

1. Solution Conditions

For the above system of heat-flow-deformation governing equations, the seepage equation, the constitutive equation, and the energy equation, the respective boundary conditions and initial values should be specified separately. That is, the boundary conditions and initial values for seepage, the boundary conditions and initial values for stress, and the boundary conditions and initial values for temperature must each be provided.

(1) Boundary Conditions for the Analytical Solution of the Mathematical Equations of Pore Fluid Seepage

Let the domain occupied by the fluid be denoted as $\Omega$, with its boundary as $\Gamma$, i.e., the initial and boundary conditions for fluid seepage in the porous medium are as follows:

$$\text{Initial value:} \quad P|_{t=0,\Omega} = \overline{P}_0 \tag{2.62}$$

$$\text{Boundary condition:} \quad \begin{cases} P|_{t,\Gamma_p} = \overline{P} \\ \frac{k}{\mu} \nabla p \cdot \vec{n}|_{t,\Gamma_v} = -\overline{q} \end{cases} \tag{2.63}$$

(2) Boundary Conditions for Solving the Temperature Field Equation

Let the domain occupied by the fluid be $\Omega$, with boundary $\Gamma$. The initial and boundary conditions for the pore fluid seepage are as follows:

$$\text{Initial value:} \quad T|_{t=0,\Omega} = \overline{T}_0 \tag{2.64}$$

$$\text{Boundary condition:} \quad \begin{cases} T|_{t,\Gamma_t} = \overline{T} \\ k(\nabla T) \cdot \vec{n}|_{t,\Gamma_q} = -\overline{q} \end{cases} \tag{2.65}$$

(3) Conditions for the Well-Posedness of This Structural Equation

Let the domain in which the reef framework exists be $\Omega$, with the domain boundary denoted as $\Gamma$, i.e., the boundary condition for the governing equation is:

$$\text{Boundary condition:} \quad \begin{cases} \overline{X}_i\Big|_{t,\Gamma_\sigma} = \sigma_{ij} n_j \\ u_i|_{t,\Gamma_u} = \overline{u}_i \end{cases} \tag{2.66}$$

Equations (2.46), (2.55), and (2.59), together with the boundary conditions (2.60)–(2.64), constitute a complete mathematical model for multiphysics coupling. Numerical methods can be employed to solve this mathematical model, yielding the distributions of temperature, seepage flow, stress fields, and material concentration. This provides theoretical guidance for practical applications in fields such as petroleum reservoir engineering and geotechnical engineering.

2. Synergistic Effect Analysis

(1) The Influence of Temperature and Pore Medium Deformation on Fluid Seepage

Through the analysis of the fluid seepage Eq. (2.44) mentioned above, it can be understood that there is a certain interaction between changes in the carbonate rock framework and the transport of fluid mass. In fact, fluid seepage is also subject to the combined influence of factors such as carbonate rock porosity and permeability. Therefore, it is necessary to conduct an in-depth study of the interrelationship between these two factors.

Permeability is a key quantitative indicator for studying the ability of fluids to flow through pore spaces. Traditionally, permeability has been regarded as a constant value in numerical reservoir simulations. However, in reality, this parameter is dynamically variable due to the deformation of the rock matrix. Therefore, conclusions drawn from conventional numerical simulation methods are often lacking in validity. It is thus necessary to analyse the relationship between these two factors. In the course of this analysis, in order to better understand the effect of rock deformation on permeability, it is assumed that the influence of other factors is minimal and can even be neglected.

Carman analysed the model and proposed the following relationship between permeability and porosity:

$$K = K_0\left(\frac{\phi}{\phi_0}\right)^3\left(\frac{1-\phi_0}{1-\phi}\right)^2 \tag{2.67}$$

When studying surface subsidence in Mexico City, Rivero conducted multiple experiments and derived an empirical formula relating permeability to porosity:

$$K = K_0\left[\frac{\phi(1-\phi_0)}{\phi_0(1-\phi)}\right]^n \tag{2.68}$$

In the equation: $\phi$ denotes the porosity of tuff; $\phi_0$ represents the initial porosity of tuff; $K$ is the permeability of tuff when the porosity is $\phi$; $K_0$ is the initial permeability of tuff when the initial porosity is $\phi_0$; $n$ is an empirical value, generally taken as 3.

(2) Influence of Fluid Seepage and Temperature on the Deformation of Porous Media

In the lithostatic equilibrium Eq. (2.57), the equation includes both the pressure gradient term and the temperature gradient term.

① The Impact of Fluid Seepage on the Deformation of Porous Media

The fluid seepage within a porous medium can, to a certain extent, influence the mean stress of the porous medium, causing the material to exhibit a state of deformation. The principal mechanism underlying this process is that when fluid pressure exists within the pores of the medium, it leads to a reduction in the mean stress, thereby generating a "buoyant force" effect that induces deformation in the porous medium.

② Influence of Temperature on the Deformation of Pore Medium

During the process of injecting high-temperature steam into the strata, the caprock is subjected to elevated temperatures and exhibits a swelling state. However, due to various constraints imposed by internal and external factors within the caprock, this swelling cannot occur freely and without limitation, resulting in the manifestation of thermal effects.

(3) Influence of Pore Matrix Deformation and Fluid Seepage on Temperature

In Eq. (2.61), it can be seen that the main components include the thermo-mechanical coupling term, the advection term, and the heat flux coupling term, which are represented respectively as $(1-\phi)\beta_{Tm}\frac{\partial \varepsilon_V}{\partial t}$, $\phi\rho_l c_{lV} V_l \cdot \nabla T$ and $(1-\phi)\beta_{Tm}\frac{\partial \varepsilon_V}{\partial t}$. In fact, in oil reservoirs, the essence of rock deformation is the transformation of a certain type of energy. During this transformation process, it is necessary to comprehensively balance both mechanical and thermal equilibria, which requires the use of the coupling energy equation to evaluate the relationship between rock deformation and temperature. This is primarily because the equation contains the coupling term $(1-\phi)\beta_{Tm}\frac{\partial \varepsilon_V}{\partial t}$, which can effectively reflect the state of rock deformation. The advection term $\phi\rho_l c_{lV} V_l \cdot \nabla T$ mainly refers to the energy difference generated during the process of energy inflow and outflow under different volumes and at different times. When the permeability of the oil reservoir is relatively high, the main mode of heat transfer is induced by fluid convection. However, to simplify the computational steps, the temperature gradient of the advection term can be calculated based on the content from the previous time step, meaning that the advection term can be interpreted as a volumetric force for computational purposes.

## 2.4 Mechanisms of Carbon Dioxide Geological Storage

This section focuses on the core topic of geological mechanisms for $CO_2$ storage, providing a systematic overview of the principles and technical logic underlying $CO_2$ storage in various geological formations. It covers mechanisms such as $CO_2$-enhanced oil recovery, $CO_2$-enhanced coalbed methane recovery, and $CO_2$ storage in saline aquifers. Understanding these storage mechanisms is essential for comprehending the key aspects of carbon capture, utilisation and storage (CCUS) technologies, which are critical for achieving large-scale carbon emission reductions.

### *2.4.1 Analysis of Carbon Dioxide-Enhanced Oil Recovery Mechanisms*

Most of China's oil reservoirs are characterised by low permeability. In order to achieve effective development, gas injection for enhanced oil recovery has attracted considerable attention. $CO_2$ flooding technology can not only improve recovery rates but also enable the storage of $CO_2$, offering significant economic and environmental benefits. As a result, it has received widespread interest in the field of petroleum development.

The mechanisms by which $CO_2$ enhances oil recovery can be mainly categorised as follows:

(1) Reducing the Viscosity of Crude Oil

After $CO_2$ dissolves in water, it forms carbonic acid, which acidifies the crude oil. This process can reduce the viscosity and flow resistance of the crude oil, increase the sweep efficiency and volume of water flooding, and thereby enhance the recovery rate of crude oil.

(2) Crude Oil Volumetric Expansion

After $CO_2$ is dissolved in crude oil, the volume of the crude oil swells by 110–120%, reducing capillary resistance and improving the mobility of the crude oil.

(3) Reduction of Interfacial Tension

After $CO_2$ dissolves in crude oil, it can reduce the interfacial tension between oil and water, enabling the residual oil to aggregate within the pores and move freely, thereby enhancing the permeability of the reservoir.

(4) Acid Digestion

After $CO_2$ dissolves in water, the water becomes acidic and can undergo chemical reactions with the geological matrix, resulting in further dissolution in water and increasing the permeability of the reservoir.

(5) Extraction of Light Components

After $CO_2$ is dissolved in crude oil, it undergoes exchange with light components, reducing the saturation of the remaining oil. $CO_2$ continues to dissolve until the crude oil reaches a dissolution-extraction equilibrium, thereby achieving the extraction of light components.

(6) Mixed-Phase Effects

When the oil reservoir pressure exceeds the minimum miscibility pressure, $CO_2$ forms a miscible phase with crude oil and with light components, thereby facilitating the oil displacement process.

$CO_2$ flooding can be categorised according to the displacement mechanism into miscible flooding, near-miscible flooding, and immiscible flooding. Based on the actual geological conditions of oil reservoirs in China, if the reservoir pressure exceeds the minimum miscibility pressure by more than 1 MPa, it is referred to as miscible flooding; if the reservoir pressure is lower than the minimum miscibility pressure but within a range of 1 MPa, it is termed near-miscible flooding; if the reservoir pressure is lower than the minimum miscibility pressure by more than 1 MPa, it is classified as immiscible flooding. In the process of miscible flooding, the main mechanism utilised is the miscibility effect, whereby $CO_2$ forms a miscible zone with light hydrocarbon components, thereby enhancing the oil displacement process. In immiscible flooding, $CO_2$ improves oil recovery primarily through mechanisms such as oil swelling, reduction of interfacial tension, and extraction, thus increasing the oil recovery factor and enhancing displacement efficiency.

### 2.4.2 Analysis of Carbon Dioxide-Enhanced Coalbed Methane Mechanisms

Gases are primarily adsorbed and stored within the micropores of coal. Under confined conditions, coalbed methane is produced following processes of adsorption, desorption, diffusion, dispersion, and migration, ultimately reaching dynamic equilibrium. The development of coalbed methane involves the processes of desorption, dispersion, and diffusion. The adsorption equilibrium for multicomponent systems can be determined using the Langmuir equation:

$$V_i = \frac{a_i b_i P_i}{1 + b_1 P_1 + b_2 P_2 + \cdots + b_n P_n} \tag{2.69}$$

In the equation, $V_i$ denotes the adsorption amount of each component in the coal seam; $a_i$ represents the maximum adsorption amount of a single component under the corresponding pressure; $P_i$ is the partial pressure of each component at adsorption equilibrium; $b_i$ is the adsorption equilibrium constant of a single component; $n$ refers to the mole fraction of the mixed gas component.

According to Eq. (2.69), when $CO_2$ is injected into the coal seam, the original dynamic equilibrium equation is disrupted. As a result, the adsorbed coalbed methane in the original coal seam undergoes desorption, which increases the diffusion and seepage rates, thereby enhancing the recovery efficiency.

The interactions between coalbed methane, $CO_2$, and coal molecules differ, resulting in varying adsorption capacities of coal seams for these gases. Research indicates that, under identical pressure conditions, the higher the boiling point and the stronger the interaction force, the greater the adsorption capacity of the coal seam. Under the same pressure, $CO_2$ has a higher boiling point than coalbed methane; therefore, coal seams exhibit a stronger adsorption capacity for $CO_2$. When $CO_2$ is injected into a coal seam, it competes with coalbed methane for adsorption sites, thereby displacing the methane from the coal seam. Furthermore, during the displacement process, the total pressure within the coal seam remains nearly constant. As $CO_2$ is continuously injected, its partial pressure increases while the partial pressure of coalbed methane decreases. Through ongoing competitive adsorption, coalbed methane is gradually displaced.

The competitive adsorption between $CO_2$ and coalbed methane is a staged process. Therefore, the $CO_2$-enhanced coalbed methane recovery method is best implemented using an injection-then-production approach. That is, after injecting $CO_2$, production is delayed for a period of time to allow the coal seam to reach a new dynamic equilibrium before commencing gas extraction. This strategy can significantly improve recovery efficiency. At the same time, compared with conventional pressure-depletion production, the $CO_2$-enhanced recovery method does not require reducing the pressure of the coal seam, thus avoiding issues associated with pressure drawdown and seepage. This not only ensures the sustained production of coalbed methane, but also extends the development period of the coal seam, further increasing both the cumulative production and recovery rate of coalbed methane.

### 2.4.3 *Analysis of Carbon Dioxide Aquifer Storage Mechanisms*

The mechanisms governing geological storage of $CO_2$ in aquifers are relatively complex, with storage capacity and efficiency jointly determined by both physical and geochemical trapping processes. Initially, $CO_2$ is impeded by the physical trapping mechanisms of the caprock overlying the reservoir structure, after which a series of geochemical reactions occur within the geological matrix and formation water. Based on the hydrodynamic and physicochemical properties of the reservoir, the trapping mechanisms can be categorised into four types: structural trapping, residual trapping, solubility trapping, and mineral trapping.

When $CO_2$ is initially injected into deep aquifers, structural trapping and residual trapping are the dominant mechanisms. Structural trapping occurs through the physical barrier provided by the caprock, which prevents the upward migration of $CO_2$, while residual trapping retains $CO_2$ within the reservoir through capillary forces associated with water flow in the formation. As the duration of storage increases, the relative contributions of these trapping mechanisms change. Once fluid pressure equilibrium is reached, the primary trapping mechanism shifts to the slow and persistent processes of $CO_2$ diffusion and dissolution at the interface between $CO_2$ and formation water. Simultaneously, as minerals dissolve, the ionic composition of the formation water changes, initiating mineral trapping processes.

(1) Structural Sealing Mechanism

The mechanism of structural geological trapping of $CO_2$ is very similar to that of oil and natural gas preservation within closed geological formations. When $CO_2$ is initially injected into the reservoir, it exists in a gaseous or supercritical fluid state under the influence of temperature and pressure. As the density of supercritical $CO_2$ is lower than that of water, buoyancy drives the upward migration of $CO_2$. The presence of an impermeable caprock layer prevents its lateral and upward movement, causing the $CO_2$ to accumulate beneath the caprock and thus forming a structural geological trap. The impermeable or low-permeability caprock above the reservoir structure acts as a barrier to the lateral and upward migration of $CO_2$ driven by buoyancy, resulting in its retention beneath the caprock. This caprock is referred to as the "seal" or "caprock". The formation of structural geological traps is influenced by variations in the types of geological seals resulting from different modes of caprock formation. Geological structures that give rise to such trapping include anticlines, faults, and stratigraphic pinch-outs. During the process of $CO_2$ geological storage, it is essential to avoid subjecting the structural formation to excessive pressure, in order to prevent fracturing of the caprock or reactivation of faults.

(2) Free Trapping Mechanism

The conditions for hydrodynamic trapping differ from those of structural and stratigraphic traps, as well as capillary seals; it is formed by hydrodynamic closure. This trapping mechanism, like structural and stratigraphic closures, begins to function immediately after $CO_2$ injection. The distinction lies in the fact that the lateral migration of $CO_2$ within a hydrodynamic closure is not impeded.

When, during the infiltration process, the hydraulic pressure of groundwater flow is opposite in direction to the buoyancy-driven migration of $CO_2$ and their magnitudes are approximately equal, this can impede and accumulate $CO_2$, resulting in hydrodynamic trapping. When $CO_2$ is injected into the deep aquifer beneath a confining layer, hydrodynamic trapping will occur. Groundwater in deep aquifers flows over relatively long timescales within a regional or basin-scale flow system. In such systems, the flow velocity of the fluid is measured in centimetres per year (cm/a), while the migration distance is calculated in units of tens of kilometres or several hundred kilometres.

If $CO_2$ is injected into such a system, although there is no distinct caprock as in structurally closed traps to act as a barrier to the lateral migration of $CO_2$, the $CO_2$ can still migrate along the dip of the strata under the influence of buoyancy, albeit at a very slow rate. It may take tens of thousands to several million years for this $CO_2$ to migrate to the shallow layers of the discharge area. During this process, other trapping mechanisms such as residual trapping, solubility trapping, and mineral trapping will also act simultaneously, ultimately resulting in no free $CO_2$ reaching the shallow subsurface. In addition, during the migration of $CO_2$, it is also possible for it to encounter structurally closed traps and become contained within them.

(3) Mechanism of Dissolution Sealing

Dissolution trapping refers to the process whereby injected $CO_2$ comes into contact with formation water in the aquifer, and, through mechanisms such as diffusion and convection, undergoes dissolution reactions, thereby becoming trapped in the aquifer in a dissolved state. This is a continuous, time-dependent process. $CO_2$ dissolves in the aquifer as aqueous phase components, namely $CO_2$ (aq), $H_2CO_3$, $HCO_3^-$, and $CO_3^{2-}$. The dissolution of $CO_2$ in water involves both physical and chemical processes. Physical dissolution refers to the process by which $CO_2$ molecules dissolve in water without any change in molecular structure, with the solubility governed by Henry's law under specific temperature and pressure conditions. Chemical dissolution refers to the reaction of $CO_2$ with water to form carbonic acid ($H_2CO_3$), which then ionises to produce $HCO_3^-$ and $H^+$. Compared with the amount of physical dissolution, the quantity of chemical dissolution is relatively small.

The greatest advantage of solubility trapping is that, once the injected $CO_2$ dissolves in water, it no longer exists as a separate phase. As a result, the buoyant force that would otherwise drive the upward migration of $CO_2$ is eliminated. During the two-phase flow of $CO_2$ and water, a portion of the $CO_2$ dissolves into the saline water before reaching the top of the reservoir. Simultaneously, the resulting mixture, being denser than the saline water, sinks, while fresh saline water rises to fill the pressure void, thereby forming a counterflow. This process increases the amount of fresh saline water that comes into contact with the $CO_2$, further enhancing the degree of dissolution and expanding the area of saline water exposed to $CO_2$. The extent of dissolution is determined by the duration and the saturation level of $CO_2$ in the formation water, while the amount and rate of $CO_2$ dissolution in the reservoir fluid mainly depend on the chemical composition of the formation water and the contact rate between $CO_2$ and unsaturated formation water; the higher the contact rate, the faster the dissolution. Over time, when the density of the $CO_2$-saturated fluid exceeds that of the surrounding unsaturated fluid by approximately 1%, the $CO_2$-saturated fluid will move downward under the influence of gravity, a phenomenon known as convection. Therefore, compared with buoyancy trapping mechanisms in geological structures, this form of trapping is more effective and offers greater storage security. The degree of dissolution mainly depends on the presence of highly permeable thick strata, especially those with high vertical permeability. The timescale for solubility trapping can reach 100–1000 years.

(4) Mechanisms of Mineral Storage

After $CO_2$ dissolves into the aquifer, it dissociates into ${CO_3}^{2-}$, which then reacts with various naturally occurring metal cations (such as $Ca^{2+}$ and $Mg^{2+}$) present in the formation water, thereby transforming into stable carbonate mineral phases. This reaction process is referred to as mineral trapping and represents the most permanent form of $CO_2$ geological storage. Geochemical reactions are the principal mechanism for $CO_2$ storage in saline aquifers. The geochemical reaction equation for the precipitation of carbonate minerals resulting from the reaction between injected $CO_2$ and metal cations in the aquifer is as follows:

$$CO_2(g) \rightarrow CO_2(aq) \tag{2.70}$$

$$CO_2(aq) + H_2O \rightarrow H_2CO_3 \tag{2.71}$$

$$H_2CO_3 \rightarrow H^+ + HCO_3^- \tag{2.72}$$

$$HCO_3^- \rightarrow H^+ + CO_3^{2-} \tag{2.73}$$

$$Ca^{2+} + CO_3^{2-} \rightarrow CaCO_3(s) \tag{2.74}$$

$$Mg^{2+} + CO_3^{2-} \rightarrow MgCO_3(s) \tag{2.75}$$

$$Ca^{2+} + Mg^{2+} + 2CO_3^{2-} \rightarrow CaMg(CO_3)_2(s) \tag{2.76}$$

The geochemical reactions of $CO_2$ in aquifers are initially represented by Eq. (2.70). The injected $CO_2$ dissolves in the aquifer, causing the pH of the system to decrease due to the formation of carbonic acid, as shown in Eq. (2.71). Subsequently, in Eq. (2.72), carbonic acid further dissociates to form $HCO_3^-$, and in Eq. (2.73), $HCO_3^-$ further dissociates to $CO_3^{2-}$. Thereafter, as described in Eqs. (2.74)–(2.76), metal cations present in the aquifer react with the previously formed $CO_3^{2-}$, resulting in the precipitation of carbonate minerals such as calcite ($CaCO_3$), magnesite ($MgCO_3$), and dolomite [$CaMg(CO_3)_2$].

The reaction rate of mineral carbonation storage depends on the chemical composition of groundwater and the rock matrix, temperature, pressure, the contact area between mineral particles and $CO_2$-bearing groundwater, as well as the fluid flow rate at the rock interface. The size of the interface is determined by the size of the mineral particles, while the flow rate is governed by the permeability of the rock, the hydraulic gradient, and the viscosity of the water. The viscosity of the water, in turn, is influenced by water temperature, mineralisation, and pressure. The effectiveness of $CO_2$ mineral carbonation storage varies significantly depending on the type of rock and the composition of the minerals. If the reservoir consists of carbonate rocks (mainly composed of calcite and dolomite), the geochemical reaction rate is very rapid; if the reservoir is sandstone with predominantly stable quartz grains, reactions generally do not occur or require a very long time to proceed.

When the system pH is below 4.0, $H_2CO_3$ is the main product of this series of geochemical reactions. When the system pH is at an intermediate value (approximately 6.0), $HCO_3^-$ predominates throughout the system. At higher pH values ($\geq$ 9.0), $HCO_3^-$ becomes the principal anion in the entire system. The formation of carbonate mineral precipitates is primarily determined by the pH of the brine; a pH above 9.0 is favourable for the mineral trapping process. However, once $CO_2$ is injected into the brine, the pH will decrease. Therefore, during the brine layer trapping process, it is necessary to increase and maintain the stability of the brine pH.

## 2.5 Section Summary

This section begins with the displacement mechanisms of $CO_2$ and the changes in reservoir parameters following $CO_2$ injection, analysing the mechanisms of $CO_2$-driven oil recovery and $CO_2$-enhanced coalbed methane extraction. Under supercritical conditions, $CO_2$ exhibits a density close to that of a liquid and a viscosity similar to that of a gas, thus possessing the advantages of high mobility, high density, and strong solvency. Therefore, the supercritical state of $CO_2$ is commonly utilised during the $CO_2$ storage process. In $CO_2$-driven oil recovery, $CO_2$ reduces the viscosity of crude oil, induces acid dissolution, and extracts light components, thereby enhancing oil recovery. In $CO_2$-enhanced coalbed methane extraction, compared with methane, coal has a stronger adsorption capacity for $CO_2$. As a result, after $CO_2$ injection, competitive adsorption occurs between $CO_2$ and methane, leading to the desorption of large quantities of methane and thus improving coalbed methane recovery. The geological conditions of the storage site directly affect the effectiveness and safety of $CO_2$ storage. These geological conditions include the stability of geological structures, the properties of the reservoir, and the stability of the caprock.

Petrophysics serves as the bridge connecting reservoir properties and elastic parameters. By analysing the mineral composition, pore structure, pore fluids, and pressure of the target reservoir rocks, it is possible to calculate parameters such as the reservoir's P-wave velocity, S-wave velocity, and density. The Gassmann equation discusses the calculation methods for various parameters involved in fluid substitution. Based on previous research, this section theoretically analyses the process of $CO_2$ saturation and derives the calculation methods for the reservoir's P-wave velocity, S-wave velocity, and density after $CO_2$ saturation. Using a specifically designed $CO_2$ injection well as an example, the changes in the reservoir's elastic parameters following $CO_2$ saturation are calculated. When $CO_2$ replaces 70% of the original oil, the change in P-wave velocity reaches − 200 m/s, providing a theoretical basis for monitoring $CO_2$ migration using time-lapse seismic data.

The multiphase and multiphysics coupling theory establishes a dynamic description framework for the sealing process: by integrating the governing equations for the seepage field, the solute transport field, the deformation field of the caprock, and the temperature field, it elucidates the interactions and coupling relationships among these fields (such as the impact of seepage on temperature distribution and the feedback of caprock deformation on permeability). This provides a comprehensive analysis of the complex multiphysical field interaction mechanisms during the sealing process.

This section covers the progression from fundamental concepts to theoretical modelling and then to application mechanisms. It addresses the key aspects essential for understanding the scientific basis and feasibility of $CO_2$ geological storage technology, providing a solid theoretical foundation for subsequent engineering design, numerical simulation, and performance evaluation.

## References


1. Krooss F, Bergen Y, Gensterblum N et al (2002) High-pressure methane and carbon dioxide adsorption on dry and moisture-equilibrated Pennsylvanian coals. Int J Coal Geol 51:69–92
2. Xu Y, Li X (2009) Factors influencing underground $CO_2$ storage and enhanced oil recovery effect. Xinjiang Petrol Geol 30(4):493–495
3. Sheng J (2006) Comprehensive mathematical model of multiphase flow in $CO_2$ flooding processes. J Petrol Mech Eng 25(9):1893–1893
4. Du Y, Chen Y, Yu H (2005) Measurement and prediction of minimum miscibility pressure in $CO_2$ flooding. Petrol Geol Recov Effic 06:64–66
5. Fitts JP, Peters CA (2013) Caprock fracture dissolution and $CO_2$ leakage. Rev Miner Geochem 77(1):459–479
6. Raziperchikolaee A, Pasumarti A, Mishra S (2020) The effect of natural fractures on $CO_2$ storage performance and oil recovery from $CO_2$ and WAG injection in an Appalachian basin reservoir. Greenh Gases Sci Technol 10(5):1098–1114
7. Raziperchikolaee S, Alvarado V, Yin S (2013) Effect of hydraulic fracturing on long-term storage of $CO_2$ in stimulated saline aquifer. Appl Energy 102:1091–1104

8. Xu X (2023) Analysis of factors influencing $CO_2$ migration in fractured deep saline aquifers. Northeast Petroleum University, Daqing
9. Peng S, Shi S, Wang J (2025) Geological storage and monitoring of carbon dioxide. Mechanical Industry Press, Beijing
10. Li D (2021) Detailed characterisation of $CO_2$ storage reservoirs and research on migration monitoring methods. China University of Mining and Technology, Beijing
11. Avseth P, Mukerji T, Mavko G (2005) Quantitative seismic interpretation: applying rock physics tools to reduce interpretation risk. Cambridge University Press
12. Wang Z (2001) Fundamentals of seismic rock physics. Geophysics 66(2):398–412
13. Hill R (1952) The elastic behaviour of crystalline aggregate. Proc Phys Soc Sect A 65(5):349–354
14. White JE (1975) Computed seismic speeds and attenuation in rocks with partial gas saturation. Geophysics 40(2):224–232
15. Mavko G (1998) Bounds on low-frequency seismic velocities in partially saturated rocks. Geophysics 63(3):918–924
16. Zhang J, Ma Z, Qu S et al (2012) Calculation of mixed-phase fluid volume modulus during fluid substitution in carbonate reservoirs. Petrol Geophys Explor 51(2):133–137
17. Baechle GT, Eberli GP, Weger RJ et al (2009) Changes in dynamic shear moduli of carbonate rocks with fluid substitution. Geophysics 74(3):E135–E147
18. Yang H (2017) Experimental and theoretical study on the mechanism of $CO_2$ flooding based on multiphase coupling. Chengdu University of Technology
19. Li N, Cheng L (2012) Study on flow-solid coupling model considering dispersion in low-permeability oil reservoirs during $CO_2$ miscible flooding. J Petrol Mech Eng 31(S1):3055–3060
20. Shi J (2015) Numerical simulation study of steam-assisted gravity drainage based on thermo-hydro-mechanical coupling model. Southwest Petroleum University
21. Ehsan Eshraghi S, Reza Rasaei M, Zendehboudi S (2016) Optimisation of miscible $CO_2$ EOR and storage using heuristic methods combined with capacitance/resistance and Gentil fractional flow models. J Nat Gas Sci Eng 32:304–318
22. Jin Z, Wang Z, Mao C (2020) Dominant mechanisms and applications of non-miscible $CO_2$ flooding in low-permeability M oil reservoirs. Oil Gas Reserv Eval Dev 10(3):68–74
23. Li Y (2020) Advances and prospects of $CO_2$ flooding technology for enhanced recovery in low-permeability oil reservoirs. Petrol Geol Recov Effic 27(1):1–10
24. Li N (2018) Numerical simulation study on $CO_2$ storage and enhanced coalbed methane recovery in deeply buried Central Ordos Basin. China University of Mining and Technology, Xuzhou
25. Connell LD, Detournay C (2009) Coupled flow and geomechanical processes during enhanced coal seam methane recovery through $CO_2$ sequestration. Int J Coal Geol 77(1):222–233
26. Ju H, Li P, Li Y et al (2023) Calculation methods and application scope for $CO_2$ storage capacity based on different trapping mechanisms in saline aquifers. J Northwest Univ (Nat Sci Ed) 53(6):913–925
27. Shu W (2020) Study on the migration patterns of carbon dioxide storage in deep saline aquifers. Dalian Maritime University, Dalian
28. Bachu S, Bonijoly D, Bradshaw J et al (2007) $CO_2$ storage capacity estimation: methodology and gaps. Int J Greenh Gas Control 1(4):430–443
29. De Silva PNK, Ranjith PG (2012) A study of methodologies for $CO_2$ storage capacity estimation of saline aquifers. Fuel 93:13–27

# Chapter 3
# Numerical Simulation Techniques for Carbon Dioxide Geological Storage

Numerical simulation technology not only enables quantitative prediction of storage effectiveness and resource development efficiency, but also facilitates the early identification of engineering risks and the optimisation of technical parameters. This provides robust technical support for the design of $CO_2$ geological storage engineering schemes, effectiveness verification, and decision-making processes. This section introduces the corresponding numerical simulation methods for $CO_2$-enhanced oil recovery ($CO_2$-EOR), $CO_2$-enhanced coalbed methane recovery ($CO_2$-ECBM), and saline aquifer storage.

## 3.1 Numerical Simulation of Carbon Dioxide-Enhanced Oil Recovery

During the process of $CO_2$ storage in oil and gas reservoirs, the continuous injection of $CO_2$ leads to an increase in pore pressure within the reservoir rocks, which in turn alters the stress state of the rocks. This results in rock deformation and changes in the pore structure, thereby affecting the storage space and flow pathways for the permeating fluids. Such changes not only impact the physical properties of the reservoir rocks, potentially causing rock damage and failure and forming high-permeability channels for $CO_2$ leakage, but may also lead to excessive stress and deformation in the caprock, resulting in fracturing and ultimately causing the failure of the $CO_2$ storage project. The flow of fluids within the reservoir and the stress–strain behaviour of the rocks are mutually influential and coupled processes. At the same time, the presence of $CO_2$ has a significant impact on the mechanical properties and stability of the rocks, which cannot be ignored. Therefore, research into the coupled mechanisms of seepage, stress, and damage during $CO_2$ storage is of great importance. On the basis of an in-depth analysis of the interactions among these three factors, this section establishes a mathematical model that characterises the coupled mechanisms



D. Li and Y. Guo, *Geological Storage and Monitoring of Carbon Dioxide*,
https://doi.org/10.1007/978-981-92-3621-3_3

of seepage, stress, and damage during $CO_2$ storage in depleted oil reservoirs. This includes the seepage equations describing multiphase flow behaviour, the constitutive equations characterising the stress–strain relationship of the rocks, the fluid–solid coupling equations, and the coupled seepage-stress-damage equations.

### 3.1.1 Seepage-Stress-Damage Coupling Mechanism

The study of the coupled processes of seepage, stress, and damage during $CO_2$ storage encompasses a wide range of disciplines, including fluid mechanics, elastoplastic mechanics, damage mechanics, and physicochemistry. The main focus is on the coupled interactions between the rock matrix framework and the migrating fluids. However, the mutual interactions among the stress field, seepage field, and chemical field render the coupling mechanisms highly complex. Therefore, the governing equations must incorporate coupling terms that can characterise the interrelationships among these fields. Moreover, any change in one field will induce corresponding changes in the intrinsic relationships and governing equations of the other fields.

The interactions between reservoir rocks and fluids can alter the mechanical properties and physicochemical characteristics of the rock framework. On the one hand, the rock may undergo argillisation and softening, and the framework grains may react with the fluids present in the reservoir, leading to processes such as dissolution/precipitation and corrosion. On the other hand, fluids can influence the mechanical properties of the rock by modifying the effective stress applied to the rock mass. The mutual interactions between the various fields are as follows:

(1) Coupled Interaction Between Seepage and Stress Fields: The influence of the seepage field on the stress field is primarily manifested in the way fluid flow alters the effective stress exerted on the rock mass framework, resulting in deformation of the rock mass. Simultaneously, the physicochemical actions of the fluid can modify the mechanical properties of the rock mass. Conversely, the effect of the stress field on the seepage field is mainly reflected in the fact that, when deformation occurs in the rock mass framework, fractures within the rock mass may open or close, thereby changing the flow pathways of the fluid and consequently affecting the permeability coefficient.
(2) Coupled Chemical–Mechanical Interactions: The influence of the chemical field on the mechanical field is primarily manifested through hydration reactions that induce dissolution or precipitation within the rock mass. This, in turn, alters the mechanical parameters of the rock mass, such as cohesion, internal friction angle, and elastic modulus. The relationship between chemical reactions and the mechanical parameters of the rock mass is typically determined using corresponding chemical kinetics calculations. Conversely, the impact of the mechanical field on the chemical field is mainly reflected in the deformation, damage, and fracturing of the rock mass caused by changes in the stress field. These

processes modify the contact area between fluids and rock, thereby affecting solute transport and, consequently, influencing the chemical field.

(3) Interaction Between Chemical and Hydrological Fields: The influence of the chemical field on the hydrological field is primarily reflected in the fact that the dissolution of minerals through chemical reactions alters the density and viscosity of the fluid, thereby affecting the flow characteristics of the fluid. Additionally, mineral dissolution can lead to changes in the pore structure and connectivity of the rock matrix, impacting its porosity and permeability, and thus influencing fluid migration. Conversely, the impact of the hydrological field on the chemical field is mainly manifested in the way fluid pressure, flow velocity, and saturation affect the rates of gas–solid dissolution as well as dissolution/precipitation reactions.

### 3.1.2 Seepage-Stress-Damage Coupled Control Equation

The coupled field control equations are primarily concerned with resolving the interactions and coupling relationships among multiple fields, clarifying the variation patterns of each field under the influence of others. This encompasses not only the changes within the individual field equations themselves, but also the description of the evolution of various intermediate physical quantities. The coupled equations established in this section are mainly based on the theories of elastoplastic mechanics, damage mechanics, seepage mechanics, and hydrochemistry. The principal governing equations include: the stress field equation, the seepage field equation, the damage evolution equation, as well as the coupled seepage–stress–damage equations.

1. Stress Field Equation

The stress field equations of oil-bearing reservoir frameworks mainly comprise the following three components:

(1) Equilibrium Equation

Under the action of total stress, the force balance equation for the solid-phase medium in an oil reservoir is as follows:

$$\sigma_{ij,j} + f_i = 0 \tag{3.1}$$

In the equation, $\sigma_{ij,j}$ denotes the total stress, and $f_i$ represents the body force. The invariant form of Eq. (3.1) is as follows:

$$\nabla \cdot \sigma + \mathrm{f} = 0 \tag{3.2}$$

Reservoir sandstone is a type of porous medium saturated with multiphase fluids. Under external loading, the total stress is jointly borne by the rock matrix and the fluids within the pore spaces. However, only the stress transmitted through the rock matrix leads to deformation of the sandstone. Based on this characteristic, Terzaghi proposed the effective stress principle in 1925. Subsequently, many scholars have revised and refined Terzaghi's principle, resulting in the modified Terzaghi principle, which is expressed as follows:

$$\sigma'_{ij} = \sigma_{ij} - \alpha\delta_{ij}P_s \tag{3.3}$$

In the equation, $\alpha$ is the Biot coefficient, with a value range of 0–1; $\delta_{ij}$ is the Kronecker delta; $P_s$ is the pore pressure. For multiphase flow, its calculation formula is:

$$P_s = \sum_{i=1}^{n_i} S_i P_i \tag{3.4}$$

Substituting Eq. (3.3) into Eq. (3.2), the force equilibrium differential equation for the reservoir rock is obtained as follows:

$$\nabla \cdot \sigma' + \nabla P_s + \mathrm{f} = 0 \tag{3.5}$$

(2) Geometric Equations

According to classical linear elasticity theory, the stress tensor is equivalent to the displacement gradient, and the geometric equation can be expressed as:

$$\varepsilon_{ij} = u_{(i,j)} \tag{3.6}$$

In the equation, $u_{(i,j)}$ can be expressed by $\frac{1}{2}\left(u_{i,j} + u_{j,i}\right)$.

(3) The Present Constitutive Equation

When establishing the stress–strain relationship of the slope body, the irreversibility of plastic deformation is taken into account. Therefore, the incremental form of the constitutive relationship is adopted. The incremental form of the elastoplastic equation is as follows:

$$\left\{d\sigma'_{ij}\right\} = [D]\left\{d\varepsilon_{ij}\right\} \tag{3.7}$$

The compliance criterion adopts the D-P criterion, which is expressed as follows:

$$f = aI_1 + \sqrt{J_2} - K = 0 \tag{3.8}$$

In the equation, $I_1$ denotes the first invariant of the stress tensor, $I_1 = \sigma_1 + \sigma_2 + \sigma_3$; $I_2$ represents the second invariant of the stress tensor, $I_2 =$

$\frac{1}{6}\left[(\sigma_1 - \sigma_2)^2 + (\sigma_2 - \sigma_3)^2 + (\sigma_3 - \sigma_1)^2\right]$, $a$ and $K$ are constants reflecting the frictional and cohesive properties of the material, respectively.

After the rock mass has been damaged, its yield condition can be rewritten as follows:

$$f^D = \alpha I_1 + \sqrt{J_2} - (1 - D)K = 0 \tag{3.9}$$

In the equation, D represents the damage variable.

2. Seepage Field Equation

(1) Infiltration Field Control Equation

The principle of $CO_2$ storage in depleted oil and gas reservoirs involves injecting compressed liquid $CO_2$ underground for storage within these reservoirs. Consequently, during the process of $CO_2$ storage and utilisation, in addition to the original reservoir fluid flow, the injection of $CO_2$ further increases the complexity of subsurface fluid movement within the reservoir. To ensure the safety of storage, reservoirs with good closure conditions and low permeability are typically selected. Therefore, when establishing the seepage equations for $CO_2$ storage in depleted oil and gas reservoirs, it is necessary to consider the multiphase flow processes in the reservoir as well as the characteristics of low-velocity non-linear seepage. In the area studied in this section, the average reservoir permeability is 18.3 mD, with some regions exhibiting permeabilities below 10 mD, classifying them as low-permeability reservoirs. Accordingly, a low-permeability non-linear seepage equation that accounts for the threshold pressure gradient is chosen to characterise fluid flow in the porous reservoir medium, namely the non-linear Darcy equation, which is expressed as follows:

$$v = -\frac{k}{\mu}\left(\frac{\partial P}{\partial x} - \lambda\right) \tag{3.10}$$

In the equation, $v$ denotes the seepage velocity; $k$ represents the permeability; $\mu$ is the fluid viscosity; and $\lambda$ refers to the threshold pressure gradient.

The continuity equation is the mathematical expression of the conservation of mass for fluids. By analysing a small control volume in a Cartesian coordinate system, the mathematical equation governing fluid seepage can be derived. Assume that the control volume has edge lengths of $\Delta x, \Delta y, \Delta z$, with its sides parallel to the coordinate axes $x, y, z$. The mass fluxes of the fluid in the $x, y, z$ directions are $\rho v_x$, $\rho v_y$, $\rho v_z$, respectively. The definition of mass conservation for fluids is as follows: within a unit time interval $\Delta t$

$$\frac{\partial(\rho V_x)}{\partial x} + \frac{\partial(\rho V_y)}{\partial y} + \frac{\partial(\rho V_z)}{\partial z} = \frac{\partial(\rho\phi)}{\partial t} \tag{3.11}$$

In the equation, $\phi$ denotes the porosity. For deformable media, Eq. (3.11) can be written as:

$$\frac{\partial(\rho V_x)}{\partial x}+\frac{\partial(\rho V_y)}{\partial y}+\frac{\partial(\rho V_z)}{\partial z}+\phi\frac{\partial\rho}{\partial t}+\rho\frac{\partial\phi}{\partial t}=0 \tag{3.12}$$

As the fluid within the pores of the rock mass is compressible, and when the effect of temperature on fluid density is neglected, the fluid density $\rho$ is a function of pressure P, namely:

$$\rho=\rho_0\left[1+C_f(P-P_0)\right] \tag{3.13}$$

In the equation, $C_f$ denotes the compressibility coefficient of the fluid, which is expressed as follows:

$$C_f=-\frac{1}{\rho}\left(\frac{\partial\rho}{\partial P}\right)_T \tag{3.14}$$

Cause

$$\phi\frac{\partial\rho}{\partial t}=\phi\rho_0 C_f\frac{\partial P}{\partial t} \tag{3.15}$$

The mass conservation equation for the solid carbonate rock framework in the reservoir can be expressed as follows:

$$\frac{\partial}{\partial t}(1-\phi)\rho_s+\nabla\cdot(1-\phi)\rho_s v_s=0 \tag{3.16}$$

In the equation, $\rho_s$ denotes the density of the rock, and $v_s$ represents the velocity of the solid skeleton. Assuming that the rock skeleton particles are incompressible and that the deformation of the reservoir rock is caused by the rearrangement of the skeleton particles, the density of the rock in the reservoir remains constant. In this case, Eq. (3.16) can be rewritten as:

$$\frac{\partial}{\partial t}(1-\phi)+\nabla\cdot(1-\phi)v_s=0 \tag{3.17}$$

Further rearrangement of Eq. (3.17) yields:

$$\frac{\partial}{\partial t}(1-\phi)+v_s\cdot\nabla(1-\phi)+(1-\phi)\nabla\cdot v_s=0 \tag{3.18}$$

Among them

$$\nabla\cdot v_s=\nabla\cdot\frac{du_s}{dt} \tag{3.19}$$

In the equation, $u_s$ represents the displacement of the solid skeleton, and its relationship with the volumetric strain of the rock can be expressed as follows:

$$\varepsilon_v = \nabla \cdot u_s \tag{3.20}$$

Substituting Eq. (3.20) into Eq. (3.19) yields:

$$\nabla \cdot v_s = \frac{d\varepsilon_v}{dt} \tag{3.21}$$

Substituting Eq. (3.21) into Eq. (3.17), we obtain:

$$\frac{d(1-\phi)}{dt} + (1-\phi)\frac{d\varepsilon_v}{dt} = 0 \tag{3.22}$$

By rearranging Eq. (3.22), we obtain:

$$-\frac{1}{1-\phi}\frac{d(1-\phi)}{dt} = \frac{d\varepsilon_v}{dt} \tag{3.23}$$

By integrating Eq. (3.23), the relationship between rock porosity and volumetric strain can be expressed as follows:

$$\phi = 1 - (1-\phi_0)e^{-\varepsilon_v} \tag{3.24}$$

In the equation, $\phi_0$ represents the initial porosity.

From Eq. (3.24), we obtain:

$$\frac{\partial \phi}{\partial t} = (1-\phi_0)e^{-\varepsilon_v}\frac{\partial \varepsilon_v}{\partial t} \tag{3.25}$$

By substituting Eqs. (3.10), (3.15), and (3.25) into Eq. (3.12), and assuming that the mass flow rate of fluid injected (or produced) per unit volume is q, the mathematical model for single-phase fluid flow in a deformable medium can be expressed as:

$$\nabla \cdot \left[\frac{\rho k}{\mu}(\nabla P - \lambda)\right] + \phi \rho_0 C_f \frac{\partial P}{\partial t} + (1-\phi_0)e^{-\varepsilon_v}\frac{\partial \varepsilon_v}{\partial t} + q = 0 \tag{3.26}$$

In view of the migration and storage characteristics of $CO_2$ in depleted oil reservoirs, and in order to establish a multiphase flow model, the following assumptions are specifically introduced:

(a) The seepage in oil reservoirs is isothermal seepage;
(b) The seepage flow in each phase conforms to the nonlinear Darcy's law;
(c) The reservoir fluid consists of two phases: oil and water, while the injected $CO_2$ phase transforms into a gas phase;
(d) $CO_2$ is soluble in both the oil phase and the aqueous phase;
(e) The reservoir sandstone is weakly compressible and isotropic in all directions.

In homogeneous and isotropic reservoirs, the equations of motion for the three phases—oil, gas, and water—are as follows:

$$v_o = -\frac{KK_{ro}}{\mu_o}(\nabla P_o - \lambda_o) \tag{3.27a}$$

$$v_w = -\frac{KK_{rw}}{\mu_w}(\nabla P_w - \lambda_w) \tag{3.27b}$$

$$v_g = -\frac{KK_{rg}}{\mu_g}(\nabla P_g - \lambda_g) \tag{3.27c}$$

In the equation, $K$ denotes the absolute permeability of the rock; $K_{ro}$, $K_{rw}$, and $K_{rg}$ represent the relative permeabilities of the oil, water, and gas phases, respectively; $\mu_o$, $\mu_w$, and $\mu_g$ are the dynamic viscosity coefficients of oil, water, and gas, respectively; $P_o$, $P_w$, and $P_g$ denote the pore pressures of the oil, water, and gas phases, respectively; $\lambda_o$, $\lambda_w$, and $\lambda_g$ represent the threshold pressure gradients for the oil, water, and gas phases, respectively.

Substituting Eqs. (3.27a), (3.27b), and (3.27c) into Eq. (3.26), the seepage flow equation for oil–gas–water three-phase flow can be obtained as follows:

Oil phase:

$$\nabla \cdot \left[\frac{\rho_o KK_{ro}}{\mu_o}(\nabla P_o - \lambda_o)\right] + \phi S_o \rho_o C_f \frac{\partial P}{\partial t} + S_o(1-\phi_0)e^{-\varepsilon_v}\frac{\partial \varepsilon_V}{\partial t} + q_o = 0 \tag{3.28a}$$

Aqueous phase:

$$\nabla \cdot \left[\frac{\rho_w KK_{rw}}{\mu_w}(\nabla P_w - \lambda_w)\right] + \phi S_w \rho_w C_f \frac{\partial P}{\partial t} + S_w(1-\phi_0)e^{-\varepsilon_v}\frac{\partial \varepsilon_V}{\partial t} + q_w = 0 \tag{3.28b}$$

Gas phase:

$$\nabla \cdot \left[\frac{\rho_g KK_{rg}}{\mu_g}(\nabla P_g - \lambda_g)\right] + \phi S_g \rho_g C_f \frac{\partial P}{\partial t} + S_g(1-\phi_0)e^{-\varepsilon_v}\frac{\partial \varepsilon_V}{\partial t} + q_g = 0 \tag{3.28c}$$

In the equation, $q_o$, $q_w$, and $q_g$ represent the injection terms for oil, water, and gas, respectively, i.e., the source/sink terms; $S_o$, $S_w$, and $S_g$ denote the saturations of the oil, water, and gas phases, respectively.

(2) Auxiliary Equations

In addition to the fundamental equations mentioned above, some auxiliary equations are also required when solving the problem:

Saturation constraint equation:

$$S_o + S_w + S_g = 1 \tag{3.29}$$

Capillary pressure equation:

$$P_{cgo} = P_g - P_o \tag{3.30a}$$

Equation for relative permeability:

$$P_{cgw} = P_g - P_w \tag{3.30b}$$

In oil–water–gas three-phase systems, the gas phase is considered a non-wetting phase. Therefore, the relative permeability of the gas phase is solely a function of gas saturation. Assuming that the relative permeability of the water phase is only affected by water saturation and that the impact of hysteresis is neglected, the permeability equations for the water and gas phases can be expressed by Eqs. (3.31a) and (3.31b), respectively. The oil phase, as the intermediate wetting phase, has its relative permeability given by Eq. (3.31c):

$$K_{rw} = K_{rw}(S_w) \tag{3.31a}$$

$$K_{rg} = K_{rg}\left(S_g\right) \tag{3.31b}$$

$$K_{ro} = \frac{(S_w - S_{wc})K_{row} + \left(S_g - S_{gr}\right)K_{rog}}{(S_w - S_{wc}) + \left(S_g - S_{gr}\right)} \tag{3.31c}$$

In the equation, $S_{wc}$ denotes the irreducible water saturation; $S_{gr}$ represents the residual gas saturation; $K_{row}$ is the relative permeability of the oil phase during the oil–water two-phase flow; and $K_{rog}$ is the relative permeability of the oil phase during the oil–gas two-phase flow.

### 3.1.3 Mechanical Damage Equation of Solid Phase Medium

During the process of $CO_2$ storage in depleted oil reservoirs, the extraction of oil and gas resources often occurs simultaneously. The injection and production process can be regarded as a loading and unloading process for the rock, during which the original open structural surfaces and microfractures within the reservoir rock gradually open or close, leading to changes in the internal structure. This process is accompanied by energy transformation and dissipation. Research has shown that energy dissipation can cause deterioration of the reservoir rock and loss of strength, which are intrinsic factors leading to rock damage and fracturing. Therefore, this section defines rock damage from the perspective of energy dissipation, establishes a fundamental theoretical framework for damage, and expresses the evolution equation for rock damage as follows:

$$D_s = 1 - \exp\left[1 - \beta\left(U^d - U_0^d\right)\right]^{\frac{1}{k}} \tag{3.32}$$

In the equation, $U^i$ represents the dissipated energy of the loess body under cyclic loading; $U_0^i$ denotes the initial dissipated energy; $\beta$ and $k$ are material parameters.

Among these, the magnitude of the dissipated energy $U^i$ is not only related to $\sigma - \varepsilon$, but is also influenced by the upper and lower limit stresses of the applied load. Its value is equal to the area enclosed by the corresponding hysteresis loop, which can be calculated using Eq. (3.33):

$$U^d = \int_{\varepsilon_1}^{\varepsilon_2} (\sigma_2 - \sigma_1) d\varepsilon \tag{3.33}$$

In the equation, $\sigma_1$, $\sigma_2$ represent the upper and lower stress limits, respectively, during a cyclic loading process.

### 3.1.4 Chemical Field and Damage Evolution Equations

(1) Mass Conservation Equation

$$\frac{\partial C_i}{\partial t} = -\upsilon \nabla C_i + \xi \nabla^2 C_i - \sum_{j=1}^{n} v_{ij} \gamma_i \tag{3.34}$$

In the equation, $C_i$ denotes the molality of the $i$th component in the solution; $\xi$ is the diffusion coefficient; $n$ represents the total number of minerals containing component $i$; $v_{ij}$ is the stoichiometric coefficient of component $i$ in mineral $j$; $\gamma_i$ is the dissolution/precipitation reaction rate of the mineral, which can be calculated using the following equation:

$$\gamma = \hat{A}_0 \frac{N}{N_0} \hat{k} \left(1 - \frac{Q}{K_{eq}}\right) \tag{3.35}$$

In the equation, $\hat{A}_0$ denotes the initial specific surface area of the mineral; $N$ represents the number of moles of mineral per unit volume; $N_0$ is the number of moles of mineral per unit volume at the initial time; $Q$ indicates the amount of substance produced by the chemical reaction; $K_{eq}$ is the chemical reaction constant.

(2) Evolutionary Pathways of Chemical Damage

During hydrochemical reactions, the dissolution and precipitation of minerals can alter the pore volume of the rock mass. Therefore, changes in porosity can be used as a damage variable to reflect the degree of rock damage. Accordingly, the chemical damage evolution equation for the rock can be expressed as follows:

$$D_c = \frac{\sum_{i=1}^{n} \int_0^t \frac{M_i \times \gamma_i}{\rho_i V_s} dt}{1 - \phi_0} \times 100\% \tag{3.36}$$

In the formula, $M_i$ represents the molar mass of mineral $i$; $\rho_i$ denotes the density of mineral $i$; $V_s$ is the volume of the rock mass.

## 3.2 Numerical Simulation of Carbon Dioxide-Enhanced Coalbed Methane Recovery

During the $CO_2$-ECBM process, complex changes occur within the coal reservoir, involving competitive adsorption of $CO_2$ and $CH_4$ in the coal matrix, fluid flow through fractures in the reservoir, variations in reservoir energy, changes in the reservoir stress state, as well as geochemical reactions and mineral dissolution/precipitation in the fracture water. Based on the extended coal matrix–mineral–fracture dual-porosity medium system physical model, a coupled mathematical model integrating adsorption–percolation–temperature–stress–chemical multiphysics can be derived for $CO_2$-ECBM simulation studies. The construction of this mathematical model requires the fulfilment of certain fundamental assumptions.

### *3.2.1 Basic Assumptions*

(1) Assumptions for Double-Porosity Media

Coal seams are composed of organic matter, inorganic minerals, and fractures. In previous numerical simulation studies, the physical model of coal seams has often been regarded as a dual-porosity medium system consisting of a matrix and fractures. The matrix comprises organic matter and minerals, containing micropores, mesopores, and some macropores, and serves as the main site for gas adsorption; fractures include cleats and some macropores, acting as the pathways for fluid migration. In the matrix–fracture dual-porosity medium system, minerals present within the coal reservoir are classified as part of the matrix, without consideration of the geochemical reactions that may occur between minerals and $CO_2$ dissolved in fracture water during the $CO_2$-ECBM process. However, the dissolution and precipitation of certain minerals during the $CO_2$-ECBM process can have a significant impact on the porosity, permeability, and geochemical environment of the coal reservoir.

Based on the actual occurrence of materials within coal reservoirs, this book extends the dual-porosity medium system of coal seams. The expanded dual-porosity medium system not only includes the coal matrix, which contains micropores, mesopores, and some macropores, as well as the reservoir fractures that comprise cleats and some macropores, but also incorporates minerals (such as feldspar) distributed between the matrix and fractures. The extended coal seam matrix–mineral–fracture dual-porosity medium system can more accurately represent the material composition and pore–fracture structural characteristics of coal seams. When applied to numerical simulations of $CO_2$-ECBM, it enables a more precise reflection of the series of changes occurring in coal reservoirs following $CO_2$ injection.

(2) Assumptions of the Mathematical Model

$CH_4$, $CO_2$, and water, among other fluids, undergo processes such as adsorption, dispersion, seepage, dissolution, and precipitation within the coal reservoir, accompanied by dynamic changes in the physical and chemical fields. The numerical simulation study of $CO_2$-ECBM requires the construction of a multi-field coupled mathematical model that incorporates gas adsorption, fluid flow, dynamic temperature, stress variation, and chemical reactions. Based on the extended coal matrix–mineral–fracture dual-porosity medium system physical model, this book proposes the following assumptions:

(a) It is assumed that the coal reservoir consists of coal matrix, minerals, and reservoir fractures. Considering that, at reservoir temperatures close to 25 °C, the dissolution reactions of carbonate minerals dominate during the $CO_2$-ECBM process, and that the main carbonate mineral present in the Nanbu 3# coal seam is calcite, it is further assumed that only calcite participates in the dissolution reactions within the 3# coal seam, while other minerals are still classified as part of the matrix.
(b) The adsorption of $CH_4$ and $CO_2$ within the coal matrix follows the extended Langmuir adsorption isotherm, and the diffusion into the coal reservoir fractures occurs via diffusion, which conforms to Fick's first law of diffusion.
(c) Water within coal seams is predominantly present as free water, with bound water as a secondary component. Small amounts of $CH_4$ and $CO_2$ within the seam may dissolve in the water in a quasi-equilibrium state under pressure. The migration of fluids and chemical ions within the fractures of the seam is primarily governed by Darcy flow, with water movement following Darcy's law. The migration of free gas is jointly controlled by Darcy's law and gas slippage effects. Gas and water interact through capillary pressure, facilitating phase exchange. The migration of dissolved gases and chemical ions within the fractures is influenced not only by Darcy flow but also by diffusion processes.
(d) Water in coal reservoir fractures can ionise to produce $H^+$ and $OH^-$. $CO_2$ dissolved in water can form $H_2CO_3$, which ionises to produce $H^+$ and $HCO_3^-$. $HCO_3^-$ can further ionise to generate $H^+$ and $CO_3^{2-}$. Chemical reactions occur

both within the fracture water of the reservoir and on the surface of feldspar minerals. The chemical interactions between ions mainly follow the law of mass action, and the dissolution/precipitation rate of feldspar is governed by reaction kinetics.

(e) The coal reservoir is considered a homogeneous, isotropic, linear elastic porous medium. The feldspar content within the reservoir is very low, so the effects of feldspar dissolution/precipitation on the elastic modulus and porosity of the reservoir can be neglected. The volumetric changes caused by feldspar dissolution/precipitation are treated as part of the deformation of the coal reservoir.

### 3.2.2 *Adsorption Field Equation*

The adsorption of $CH_4$ and $CO_2$ within the coal matrix can be described using the extended Langmuir equation:

$$V_{Ti} = \frac{V_{Li} P_{mi}}{P_{Li}\left(1 + \sum_{i=1}^{2} \frac{P_{mi}}{P_{Li}}\right)} \exp\left[-\frac{d_2(T - T_0)}{1 + d_1 P_m}\right] \tag{3.37}$$

In the formula, $i$ denotes the gas component, $i = 1$ represents $CH_4$, and $i = 2$ represents $CO_2$; $V_{Ti}$ indicates the adsorption amount of gas component $i$ by the coal matrix, in $m^3$/t; $V_{Li}$ denotes the Langmuir adsorption volume of gas component $i$, in $m^3$/t; $P_{Li}$ represents the Langmuir adsorption pressure of gas component $i$, in MPa; $P_{mi}$ indicates the partial pressure of gas component $i$ within the coal matrix, in MPa; $T$ denotes the coal reservoir temperature, in K; $T_0$ represents the initial temperature of the coal reservoir, in K; $d_1$ indicates the pressure coefficient, $MPa^{-1}$; $d_2$ denotes the temperature coefficient, in $K^{-1}$; $P_m = P_{m1} + P_{m2}{}^{-1}$ represents the total gas pressure within the coal matrix, in MPa. The adsorbed gas mass within the coal matrix can be defined as:

$$m_i = c_{STP,i} V_{Ti} \rho_s \tag{3.38}$$

In the equation, $\rho_s$ denotes the density of the coal matrix, kg/$m^3$; $c_{STP,i}$ represents the density of the gas component $i$ in the coal matrix under standard conditions, kg/$m^3$, which can be expressed as:

$$c_{STP,i} = \frac{M_{gi} P_{STP}}{R T_{STP}} \tag{3.39}$$

In the equation, $M_{gi}$ denotes the molar mass of gas component $i$, kg/mol; $P_{STP}$ represents the pressure of the free gas under standard conditions, MPa; $R$ is the universal gas constant, 8.314510 J/(mol K); $T_{STP}$ indicates the temperature of the free gas under standard conditions, K.

Gas enters the reservoir fractures from the coal matrix via diffusion. The diffusion process conforms to Fick's first law, whereby the rate of gas diffusion is directly proportional to the concentration gradient of the gas, and can be expressed as follows:

$$q_i = -\sigma D_{Ti}\left(c_i - c_{fi}\right) \tag{3.40}$$

In the equation, $c_i$ denotes the concentration (density) of gas component $i$ within the coal matrix, kg/m$^3$; $c_{fi}$ represents the concentration (density) of gas component $i$ within the coal seam fractures, kg/m$^3$; $D_{Ti}$ indicates the diffusion coefficient of gas component $i$ in the coal matrix, m$^2$/s; σ denotes the shape factor of the coal matrix, m$^{-2}$; $\sigma D_{Ti}$ can be defined as:

$$\begin{cases} \sigma D_{Ti} = \frac{1}{\tau_{Ti}} \\ \sigma D_{T_0 i} = \frac{1}{\tau_{T_0 i}} \end{cases} \tag{3.41}$$

In the equation, $\tau_{Ti}$ denotes the adsorption time of gas component $i$, d; $\tau_{T_0 i}$ represents the adsorption time constant of gas component $i$, d.

The dispersion coefficient is closely related to the ambient temperature and can be defined as:

$$\begin{cases} D_{Ti} = D_{0i}\exp\left(-\frac{Q_i}{RT}\right) \\ D_{T_0 i} = D_{0i}\exp\left(-\frac{Q_i}{RT_0}\right) \end{cases} \tag{3.42}$$

In the equation, $Q_i$ denotes the diffusion activation energy of gas component $i$, in J/mol.

The variation in the concentration of gas adsorbed within the coal matrix is equal to the mass of gas transferred between the matrix and the fractures, namely:

$$\frac{\partial m_i}{\partial t} = q_i \tag{3.43}$$

By combining Eqs. (3.37)–(3.43) and simplifying, we obtain:

$$\begin{cases} \left\{\dfrac{V_{T1}d_1d_2(T-T_0)}{(1+d_1P_m)^2}+\dfrac{V_{L1}(P_{L2}+P_{m2})}{P_{L1}P_{L2}\left(1+\sum_{i=1}^{2}\frac{P_{mi}}{P_{Li}}\right)^2}\exp\left[-\dfrac{d_2(T-T_0)}{1+d_1P_m}\right]\right\}\dfrac{\partial P_{m1}}{\partial t} \\ +V_{T1}\left[\dfrac{d_1d_2(T-T_0)}{(1+d_1P_m)^2}-\dfrac{1}{P_{L2}\left(1+\sum_{i=1}^{2}\frac{P_{mi}}{P_{Li}}\right)^2}\right]\dfrac{\partial P_{m2}}{\partial t}-\dfrac{V_{T1}d_2}{1+d_1P_m}\dfrac{\partial T}{\partial t} \\ =-\dfrac{T_s(P_{m1}-P_{f1})}{\tau_{T_01}TP_{STP}\rho_s}\exp\left[\dfrac{Q_1}{R}\left(\dfrac{1}{T_0}-\dfrac{1}{T}\right)\right] \\ \left\{\dfrac{V_{T2}d_1d_2(T-T_0)}{(1+d_1P_m)^2}+\dfrac{V_{L2}(P_{L1}+P_{m1})}{P_{L1}P_{L2}\left(1+\sum_{i=1}^{2}\frac{P_{mi}}{P_{Li}}\right)^2}\exp\left[-\dfrac{d_2(T-T_0)}{1+d_1P_m}\right]\right\}\dfrac{\partial P_{m2}}{\partial t} \\ +V_{T2}\left[\dfrac{d_1d_2(T-T_0)}{(1+d_1P_m)^2}-\dfrac{1}{P_{L1}\left(1+\sum_{i=1}^{2}\frac{P_{mi}}{P_{Li}}\right)^2}\right]\dfrac{\partial P_{m1}}{\partial t}-\dfrac{V_{T2}d_2}{1+d_1P_m}\dfrac{\partial T}{\partial t} \\ =-\dfrac{T_s(P_{m2}-P_{f2})}{\tau_{T_02}TP_{STP}\rho_s}\exp\left[\dfrac{Q_2}{R}\left(\dfrac{1}{T_0}-\dfrac{1}{T}\right)\right] \end{cases} \tag{3.44}$$

### 3.2.3 Seepage Field Equation

The migration of $CO_2$, $CH_4$, and water within coal seam fractures follows the law of mass conservation. Based on the occurrence states of $CO_2$ and $CH_4$ within the fractures (i.e., free in the fractures and dissolved in the fracture water), as well as the occurrence state of water within the fractures, the migration equations for gases and water in coal seam fractures can be defined as follows:

$$\begin{cases} \frac{\partial(\rho_{gi}S_g\phi)}{\partial t}+\nabla\cdot\left(\rho_{gi}v_{gi}\right)+\frac{\partial\left(\rho_{dgi}S_w\phi\right)}{\partial t}+\nabla\cdot\left(\rho_{dgi}v_w-\tau\phi S_wD_{dgi}\nabla\rho_{dgi}\right)=Q_{gi} \\ \frac{\partial(\rho_wS_w\phi)}{\partial t}+\nabla\cdot(\rho_wv_w)=Q_w \end{cases} \tag{3.45}$$

In the equation, $g$ and $w$ denote gas and water, respectively; $\rho$ represents the density of the fluid within the reservoir fractures, kg/m$^3$; $v$ denotes the velocity of fluid migration within the reservoir fractures, m/s; $S$ represents the saturation of the fluid within the reservoir fractures; $\phi$ denotes the porosity of the reservoir fractures; $\tau=\phi^{1/3}S_w^{7/3}$ represents the tortuosity of the reservoir fractures; $\rho_{dgi}$ and $D_{dgi}$ denote the density and diffusion coefficient, respectively, of the gas component $i$ dissolved in water, with units of kg/m$^3$ and m$^2$/s, respectively; $Q$ is the source-sink term, kg/(m$^3$ s).

The migration of fluids within reservoir fractures follows Darcy's law. At the same time, the migration of gases is also influenced by gas slippage effects. Considering the dynamic changes in saturation of gas and water during the migration process, relative permeability is introduced. The fluid migration velocity within reservoir fractures is thus given by:

$$\begin{cases} v_{gi} = -\frac{kk_{rg}}{\mu_{gi}}\left(1 + \frac{b_k}{P_{fi}}\right)\nabla P_{fi} \\ v_w = -\frac{kk_{rw}}{\mu_w}\nabla P_w \end{cases} \tag{3.46}$$

In the equation, $k$ denotes the absolute permeability of the coal reservoir fracture, mD; $k_r$ represents the relative permeability of the fluid; $\mu$ indicates the dynamic viscosity of the fluid, Pa s; $b_k$ is the gas slippage factor, MPa; $P_{fi}$ is the pressure of the free gas component $i$ within the fracture, MPa; $P_w$ is the pressure of water within the fracture, MPa. Gas and water undergo competitive displacement within the fracture $P_w$

$$\begin{cases} P_w = P_{f1} + P_{f2} - P_c \\ P_c = P_e\left(\frac{S_w - S_{wr}}{1 - S_{gr} - S_{wr}}\right)^{-1/\lambda} \\ k_{rg} = \left(\frac{S_g - S_{gr}}{1 - S_{gr} - S_{wr}}\right)^2\left[1 - \left(\frac{S_g - S_{gr}}{1 - S_{gr} - S_{wr}}\right)^{\frac{2+\lambda}{\lambda}}\right] \\ k_{rw} = \left(\frac{S_w - S_{wr}}{1 - S_{gr} - S_{wr}}\right)^{\frac{2+3\lambda}{\lambda}} \end{cases} \tag{3.47}$$

In the equation, $P_c$ denotes the capillary pressure, MPa; $P_e$ represents the displacement pressure, MPa; $S_{wr}$ and $S_{gr}$ denote the residual gas and irreducible water saturation within the fractures, respectively; $\lambda$ indicates the pore size distribution index.

The concentration of dissolved gases in water is influenced by the partial pressure of the gas and its solubility coefficient, while the diffusion coefficient of a gas in water is determined by temperature. These can be defined as follows:

$$\begin{cases} \rho_{dgi} = c_0 P_{fi} K_{dgi} M_{gi}/P_{STP} \\ D_{dgi} = \exp\left(c_{Ti} - \frac{T_{ki}}{T}\right) \end{cases} \tag{3.48}$$

In the equation, $K_{dgi} = 10^{h_{i1}+h_{i2}+h_{i3}/T+h_{i4}\log T+h_{i5}/T^2}$ represents the solubility coefficient of the gas in water; $h_i$ is a constant related to the gas component $i$; $c_{Ti}$ is the exponential constant for the gas component $i$; $T_{ki}$ is the temperature constant for the gas component $i$, in K.

The gas and water present in coal seam fractures primarily originate from the coal matrix and geochemical reactions. Specifically, $CH_4$ is derived from the desorption of the coal matrix, while $CO_2$ originates both from the desorption of the coal matrix and from geochemical reactions between $H^+$ in water and $HCO_3^-$. Water, in turn, is produced through geochemical reactions between $H^+$ and $HCO_3^-$, as well as between

$H^+$ and $OH^-$. On this basis, the source-sink phases in the differential equations for $CH_4$, $CO_2$, and water can be expressed as follows:

$$\begin{cases} Q_{g1} = \frac{M_{g1}}{\tau T_{01} RT}\left(P_{m1} - P_{f1}\right)\exp\left[\frac{Q_1}{R}\left(\frac{1}{T_0} - \frac{1}{T}\right)\right] \\ Q_{g2} = \frac{M_{g2}}{\tau T_{02} RT}\left(P_{m2} - P_{f2}\right)\exp\left[\frac{Q_2}{R}\left(\frac{1}{T_0} - \frac{1}{T}\right)\right] - r_{H_2CO_3} M_{g2} \\ Q_w = -\left(r_{H_2CO_3} + r_{H_2O}\right) M_w \end{cases} \tag{3.49}$$

In the equation, $r$ denotes the global geochemical reaction rate, mol/(m$^3$ s); $M_w$ represents the molar mass of water, kg/mol. By combining Eqs. (3.45)–(3.49) and simplifying, we obtain:

$$\begin{cases} \frac{\partial(\rho_{g1} S_g \phi)}{\partial t} + \nabla \cdot \left(-\frac{\rho_{g1} k k_{rg}}{\mu_{g1}}\left(1 + \frac{b_k}{P_{f1}}\right)\nabla P_{f1}\right) + \frac{1}{P_{STP}} \frac{\partial(S_w \phi c_0 P_{f1} K_{dg1} M_{g1})}{\partial t} \\ + \nabla \cdot \left[\frac{c_0 P_{f1} K_{dg1} M_{g1} v_w}{P_{STP}} - \frac{c_0 \tau \phi S_w}{P_{STP}} \exp\left(c_{T1} - \frac{T_{k1}}{T}\right)\nabla\left(P_{f1} K_{dg1} M_{g1}\right)\right] \\ = \frac{M_{g1}}{\tau_{T_01} RT}\left(P_{m1} - P_{f1}\right)\exp\left[\frac{Q_1}{R}\left(\frac{1}{T_0} - \frac{1}{T}\right)\right] \\ \frac{\partial(\rho_{g2} S_g \phi)}{\partial t} + \nabla \cdot \left(-\frac{\rho_{g2} k k_{rg}}{\mu_{g2}}\left(1 + \frac{b_k}{P_{f2}}\right)\nabla P_{f2}\right) + \frac{1}{P_{STP}} \frac{\partial(S_w \phi c_0 P_{f2} K_{dg2} M_{g2})}{\partial t} \\ + \nabla \cdot \left[\frac{c_0 P_{f2} K_{dg2} M_{g2} v_w}{P_{STP}} - \frac{c_0 \tau \phi S_w}{P_{STP}} \exp\left(c_{T2} - \frac{T_{k2}}{T}\right)\nabla\left(P_{f2} K_{dg2} M_{g2}\right)\right] \\ = \frac{M_{g2}}{\tau_{T_02} RT}\left(P_{m2} - P_{f2}\right)\exp\left[\frac{Q_2}{R}\left(\frac{1}{T_0} - \frac{1}{T}\right)\right] - r_{H_2CO_3} M_{g2} \\ \frac{\partial(\rho_w S_w \phi)}{\partial t} + \nabla \cdot \left(-\frac{\rho_w k k_{rw}}{\mu_w}\nabla P_w\right) = -\left(r_{H_2CO_3} + r_{H_2O}\right) M_w \end{cases} \tag{3.50}$$

### 3.2.4 Temperature Field Equation

During the $CO_2$-ECBM process, the temperature of the coal reservoir and the internal fluids undergoes dynamic changes, accompanied by fluid adsorption/desorption, diffusion, and seepage within the coal seam. The main factors causing temperature variations in the coal reservoir and fluids include gas adsorption/desorption in the coal matrix, elastic deformation of the coal seam, heat conduction, heat exchange with the fluid, as well as the heat released during gas exsolution from water and mineral dissolution reactions. According to the law of conservation of energy, the governing equation for heat transfer in the coal reservoir and internal fluids can be defined as follows:

$$\frac{\partial\left[(\rho C_p)_{eff}T\right]}{\partial t}+\eta_{eff}\nabla T-\nabla\cdot\left(\lambda_{eff}\nabla T\right)$$
$$=K\alpha_T T\frac{\partial\varepsilon_v}{\partial t}+\sum_{i=1}^{2}q_{sti}\frac{\rho_s\rho_{gi}}{M_{gi}}\frac{\partial V_{Ti}}{\partial t}+\sum_{i=1}^{2}\frac{e_{gi}\rho_{dgi}}{M_{gi}}+\frac{k_m}{(1-\phi)}\frac{\partial\varepsilon_m}{\partial t} \tag{3.51}$$

In the equation, $(\rho C_p)_{eff}$ denotes the effective specific heat capacity of the coal reservoir containing fluid, J/(m$^3$ K); $\eta_{eff}$ is the effective heat convection coefficient, J/(m$^2$ s); $\lambda_{eff}$ represents the effective thermal conductivity, W/(m K); $\varepsilon_v$ is the volumetric strain of the coal matrix; $q_{sti}$ is the adsorption heat of gas component $i$, J/mol; $e_{gi}$ is the dissolution heat of gas component $i$ dissolved in water, J/mol; $k_m$ is the reaction heat of dawsonite, J/mol; $\varepsilon_m$ is the molar content of dawsonite in the solid part of the reservoir, mol/$\text{m}_{\text{solid}}{}^3$. The effective specific heat capacity, heat convection coefficient, and thermal conductivity can be defined as follows:

$$\begin{cases}(\rho C_p)_{eff}=(1-\phi)\rho_s C_s+\sum\limits_{i=1}^{2}\phi S_g\rho_{gi}C_{gi}+\phi S_w\rho_w C_w\\ \eta_{eff}=-\sum\limits_{i=1}^{2}\frac{\rho_{gi}C_{gi}kk_{rg}}{\mu_{gi}}\left(1+\frac{b_k}{P_{fi}}\right)\left(\nabla P_{fi}-\rho_{gi}g\right)-\rho_w C_w\frac{kk_{rw}}{\mu_w}\nabla P_w\left(\nabla P_w-\rho_{gi}g\right)\\ \lambda_{eff}=(1-\phi)\lambda_s+\phi\left(S_g\lambda_g+S_w\lambda_w\right)\\ \lambda_g=\frac{1}{2}\left[\sum\limits_{i=1}^{2}\chi_{gi}\lambda_{gi}+\left(\sum\limits_{i=1}^{2}\chi_{gi}/\lambda_{gi}\right)^{-1}\right]\end{cases} \tag{3.52}$$

In the equation, $C_s$, $C_{gi}$, and $C_w$ represent the specific heat capacities of the coal matrix, gas component $i$, and water, respectively, in J/(kg K); $\lambda_s$, $\lambda_{gi}$ and $\lambda_w$ denote the thermal conductivities of the coal matrix, gas component $i$, and water, respectively, in W/(m K); $\chi_{gi}$ is the molar fraction of gas component $i$ in the mixed gas.

### 3.2.5 Stress Field Equation

Due to the effects of factors such as effective stress, changes in reservoir temperature, and coal matrix gas adsorption/desorption, coal seams undergo certain deformations during the $CO_2$-ECBM process. Based on the theory of poroelasticity, the stress–strain relationship of fluid-bearing coal reservoirs can be expressed as follows:

$$\sigma_{kl}=2G\varepsilon_{kl}+\frac{2Gv}{1-2v}\varepsilon_{cc}\sigma_{kl}-\alpha P_f\delta_{kl}-K\alpha_T T\delta_{kl}-K\varepsilon_s\delta_{kl}-K\varepsilon_c\delta_{kl} \tag{3.53}$$

$$\begin{cases} P_f = S_w P_w + \sum_{i=1}^{2} S_g P_{fi} \\ \varepsilon_s = \sum_{i=1}^{2} \alpha_s V_{Ti} \\ \varepsilon_c = \frac{\varepsilon_m M_{vm}}{1-\phi} \end{cases} \tag{3.54}$$

In the equation, $\sigma_{kl}$ denotes the total stress, MPa; $\varepsilon_{kl}$ is the strain; $\delta_{kl}$ represents the Kronecker delta; $G = E/(2+2v)$ and $K = E/(3-6v)$ are the shear modulus and bulk modulus of coal, respectively, MPa; $\alpha = 1 - K/K_s$ is the Biot effective stress coefficient of the reservoir fracture; $K_s$ is the bulk modulus of the coal matrix, MPa; $\alpha_s$ is the gas adsorption-induced swelling strain, kg/m$^3$; $M_{vm}$ is the molar volume of the ideal gas, m$^3$/mol.

The coal seam is in a state of dynamic equilibrium, and the stress equilibrium equation as well as the strain–displacement equation can be expressed as follows:

$$\begin{cases} \sigma_{kl,l} + F_k = 0 \\ \varepsilon_{kl} = \frac{1}{2}(u_{kl} + u_{lk}) \end{cases} \tag{3.55}$$

In the equation, $F_k$ denotes the $k$ directional volumetric force component (MPa); $u_{kl}$ represents the $k$ directional displacement component; $k, l = x, y, z$.

By substituting Eqs. (3.53) and (3.54) into Eq. (3.55), the modified Navier coal seam deformation displacement equation can be derived as follows:

$$\begin{aligned} u_{k,ll} + \frac{G}{1-2v} u_{l,lk} + F_k &= \alpha\left(S_w P_{w,k} + \sum_{i=1}^{2} S_g P_{fi,k}\right) \\ &+ K\alpha_T T_k + K\left(\sum_{i=1}^{2} \alpha_s V_{Ti}\right)_k + K\frac{\varepsilon_m M_{vm}}{1-\phi_k} \end{aligned} \tag{3.56}$$

### 3.2.6 Chemical Field Equations

During the $CO_2$-ECBM process, $CO_2$ dissolves in the fracture water of the coal reservoir to form dissolved $CO_2$ ($H_2CO_3$). The dissolved $CO_2$, water, and feldspar within the fractures can undergo the following chemical reactions:

$$\begin{cases} H_2CO_3 \rightleftharpoons H^+ + HCO_3^- \\ HCO_3^- = H^+ + CO_3^{2-} \\ H_2O \rightleftharpoons H^+ + OH^- \\ CaCO_3(calcite) \rightleftharpoons CO_3^{2-} + Ca^{2+} \end{cases} \tag{3.57}$$

According to the law of mass conservation, the transport of ions in fracture water can be expressed as follows:

$$\frac{\partial\left(C_j S_w \phi\right)}{\partial t} + \nabla \cdot \left(C_j v_w - \tau \phi S_w D_j \nabla C_j\right) = Q_j \tag{3.58}$$

In the equation, $j$ denotes the ionic species in water, while $j = 1, 2, 3, 4, 5$ represents $H^+$, $HCO_3{}^-$, $CO_3{}^{2-}$, $OH^-$, and $Ca^{2+}$, respectively; $C_j$ is the concentration of ionic species $j$, in mol/L; $D_j$ is the diffusion coefficient of ionic species $j$ in water, in $m^2/s$; $Q_j$ denotes the source-sink term of chemical reactions, which equals the difference between the production and consumption rates of ionic species $j$ during geochemical reaction processes, in mol/(Ls).

According to the law of mass action, the rate of material consumption resulting from the reaction between chemical ions can be expressed as:

$$r = k_c\left(\prod_{j \in react} C_j^{-v_j} - \prod_{j \in prod} C_j^{v_j}\right) \tag{3.59}$$

In the equation, $k_c$ denotes the chemical reaction rate, mol/(Ls); $v_j$ represents the ionic strength. In addition, the dissolution rate of feldspar in water is equivalent to its molar volume change in the coal reservoir. For this study, the chemical reaction rate in Eq. (3.57) can be expressed separately as follows:

$$\begin{cases} r_{H_2CO_3} = k_c\left(\prod_{j=1}^{2} \frac{c_j}{c_0} - \frac{K_{dg2} P_{f2}}{K_{H_2CO_3} P_{STP}}\right) \\ r_{HCO_3^-} = k_c\left(\frac{c_1 c_3}{c_0 c_2} - K_{HCO_3^-}\right) \\ r_{H_2O} = k_c\left(\frac{c_1 c_4}{c_0^2} - K_{H_2O}\right) \\ r_{CaCO_3} = \frac{\partial}{\partial t}\left(\frac{\varepsilon_m}{1-\phi}\right) \end{cases} \tag{3.60}$$

In the equation, $K = 10^{h_1 + h_2 T + h_3/T + h_4 \log T + h_5/T^2}$ is the equilibrium constant of the reaction; $h_1$, $h_2$, $h_3$, $h_4$, and $h_5$ are coefficients related to specific geochemical reactions. According to chemical reaction kinetics, the variation in feldspar content within coal seams can be defined as follows:

$$\frac{\partial \varepsilon_m}{\partial t} - k_{rm} A\left|1 - \Omega^\theta\right|^\eta \operatorname{sgn}\left(1 - \Omega^\theta\right)(1 - \phi)^{-1} = 0 \tag{3.61}$$

$$\begin{cases} k_{rm} = k_{298.15}^n \exp\left[-\frac{E_n}{R}\left(\frac{1}{T} - \frac{1}{298.15}\right)\right] + k_{298.15}^a \exp\left[-\frac{E_a}{R}\left(\frac{1}{T} - \frac{1}{298.15}\right)\right] a_H^n + k_s \\ \Omega = \frac{C_3 C_5}{K_{CaCO_3}} \\ A = A_m \rho_m \varepsilon_m M_{vm}(1 - \phi) \end{cases} \tag{3.62}$$

In the equation, $k_{rm}$ denotes the global chemical reaction rate constant, mol/(m$^2$ s); $n$ and $a$ represent the neutral and acidic environments of the coal seam reservoir fracture water, respectively; $k_{298.15}$ indicates the chemical reaction rate constant at 298.15 K, mol/(m$^2$ s); $E$ denotes the activation energy of the reaction, J/mol; $k_s$ represents the chemical reaction rate constant arising from other reaction mechanisms, mol/(m$^2$ s); $K_{CaCO_3}$ denotes the equilibrium constant for the feldspar dissolution reaction, i.e., the solubility product of feldspar; $\theta$ and $\eta$ are empirical constants; $a_H^n$ represents the activity coefficient of feldspar under acidic conditions; $\text{sgn}\left(1-\Omega^{\theta}\right)$ ensures the directionality of the chemical reaction; $A_m$ is the specific surface area of feldspar, m$^2$/kg.

### 3.2.7 Darcy Combined Equation

The coal reservoir is a multiporous medium. During the storage and migration of $CH_4$, $CO_2$, and water within the reservoir, the porosity and permeability of the reservoir exhibit dynamic changes. As gas undergoes adsorption/desorption within the coal matrix, the reservoir pressure varies dynamically, temperature fluctuations occur, and mineral dissolution and precipitation take place, the pore volume of the coal seam also changes dynamically. This can be specifically expressed as follows:

$$\begin{cases} \phi = \frac{1}{1+S}[(1+S_0)\phi_0 + \alpha(S - S_0)] \\ S_0 = \frac{P_{f0}}{K_s} - \varepsilon_{s0} - \alpha_T T_0 - \frac{\varepsilon_{m0} M_{vm}}{1-\phi_0} \\ S = \varepsilon_v + \frac{P_f}{K_s} - \varepsilon_s - \alpha_T T - \frac{\varepsilon_m M_{vm}}{1-\phi} \end{cases} \tag{3.63}$$

In the equation, $\phi_0$ represents the initial porosity of the coal reservoir; $\varepsilon_{s0}$ denotes the initial strain of the coal matrix induced by gas adsorption; $\varepsilon_{m0}$ is the initial content of soluble minerals in the coal reservoir, and $mol/m^3_{solid}$.

The permeability of coal seams and porosity follow a power law relationship, which can be expressed as:

$$k = k_0 \left(\frac{\phi}{\phi_0}\right)^3 \tag{3.64}$$

In the equation, $k_0$ represents the initial permeability of the coal reservoir, in mD.

## 3.3 Numerical Simulation of Carbon Dioxide Aquifer Storage

### 3.3.1 Basic Assumptions

In order to construct a multiphysics-coupled mathematical model for the process of $CO_2$ storage in aquifers, it is necessary to appropriately simplify the complex physical and chemical interactions among $CO_2$, brine, and the rock matrix, and to propose reasonable modelling assumptions based on the underlying physico-chemical mechanisms. Firstly, prior to $CO_2$ injection, the aquifer reservoir is in an initial stable state governed by the natural stress field, subject to the combined effects of lithostatic pressure and pore pressure. When a large volume of $CO_2$ is injected into the aquifer reservoir system at a relatively high rate, the extremely low permeability of the caprock causes the pore pressure in the reservoir to rise rapidly, leading to local deformation of the reservoir framework, and potentially inducing caprock instability or fault slip. These deformation processes, in turn, affect the distribution and evolution of fluid pressure within the pore medium. Therefore, based on a comprehensive consideration of the relevant physical processes, the following fundamental assumptions are proposed to enable a reasonable simulation of the multiphysics-coupled behaviour:

(1) $CO_2$ is regarded as an ideal gas, and the effects of temperature variations on the dynamic viscosity of $CO_2$ and brine are neglected; fluid flow within the reservoir is assumed to obey Darcy's law, and dispersion behaviour conforms to Fick's first law.
(2) In the storage formation, the water is predominantly free water, with a small amount of bound water present; under high-pressure conditions, a portion of the $CO_2$ dissolves in the formation water in a quasi-stable state. Fluid migration is mainly driven by percolation flow, with the movement of $CO_2$ jointly controlled by Darcy flow and slip effects. $CO_2$ and water interact through capillary pressure, enabling mutual displacement, and their relative permeability dynamically adjusts in response to changes in saturation.
(3) The carbonate rock matrix is regarded as a homogeneous, isotropic, linear elastic continuous porous medium in all directions. During the $CO_2$ injection process, it remains in the elastic deformation stage, and its mechanical response conforms to the generalised Hooke's law, being governed solely by effective stress. Plastic deformation or irreversible nonlinear effects are not considered.

### 3.3.2 Control Equation

This book considers heat transfer, fluid flow, and deformation in fully or partially saturated porous media. A brief overview of the governing equations is provided below.

(1) Multiphase Percolation Process

Consider the general situation of flow in deformable porous media. According to the classical Darcy's law, the fluid flux $\boldsymbol{q}_f$ is defined as:

$$\boldsymbol{q}_f^\gamma = -nS^\gamma\left(\frac{k_{rel}^\gamma \boldsymbol{k}}{\mu^\gamma}(\nabla p^\gamma - \rho^\gamma \boldsymbol{g})\right) \quad (3.65)$$

where $S^\gamma$ is the degree of saturation, $p$ is the pressure, $n$ is the effective porosity of the porous medium, $\mu$ is the fluid viscosity, $k$ is the intrinsic permeability tensor, $k_{rel}^\gamma$ is the relative permeability, and $\boldsymbol{g}$ is the gravitational acceleration.

$$\frac{\partial(nS^\gamma \rho^\gamma)}{\partial t} + \nabla \cdot q_f^\gamma + \nabla \cdot (nS^\gamma \rho^\gamma \boldsymbol{u}) = Q_f^\gamma \quad (3.66)$$

For any point within the domain $x \in \Omega$, $Q_f^\gamma$ is the source term. The unknown functions in the equation are the fluid phase relative saturation $S^\gamma$ and the fluid phase pressure $\rho^\gamma$. The mass conservation equation for the fluid, combined with the deformation of the porous medium, as in the third term of the equation, where $\boldsymbol{u}$ denotes the solid displacement.

Based on the intrinsic structural equation involving fluid density, capillary pressure, and relative permeability, the equilibrium formula is derived. The boundary conditions for this problem can be either of the Neumann type or the Dirichlet type:

$$\boldsymbol{q}_f^\gamma \cdot \boldsymbol{n} = q_\Gamma^\gamma, \forall x \in \partial\Omega \quad (3.67)$$

$$p^\gamma = p_\Gamma^\gamma, S^\gamma = S_\Gamma^\gamma, \forall x \in \partial\Omega \quad (3.68)$$

This type of equation can be solved using the initial conditions of the corresponding unknowns.

(2) Deformation Process

Assuming that the solid particles are incompressible, the deformation within the porous medium can be described by the momentum balance equation for the stress term:

$$\nabla \cdot \left(\sigma - \sum_{\gamma}^{phase} S^\gamma p^\gamma \boldsymbol{I}\right) + \rho \boldsymbol{g} = 0 \quad (3.69)$$

Here, $\sigma$ denotes the effective stress of the porous medium, and $\boldsymbol{I}$ represents the constant tension.

This book adopts the conventional notation for stress and fluid pressure. The density of the porous medium is:

$$\rho = n \sum_{\gamma}^{phase} \rho^{\gamma} + (1-n)\rho^{s} \tag{3.70}$$

The components of the convective contribution and the solid-state contribution.

The fundamental constitutive equations of stress–strain behaviour are used to solve for the main variables such as displacement:

$$\sigma = \boldsymbol{E}\varepsilon \tag{3.71}$$

$$\varepsilon = \frac{1}{2}\left(\nabla \boldsymbol{u} + (\nabla \boldsymbol{u})^{T}\right) \tag{3.72}$$

In the equation, $\boldsymbol{E}$ denotes the elastic material's stress, and $\varepsilon$ represents the applied stress.

(3) Heat Transfer Process

For the issue of heat transfer, both steady flow and convective flux are considered. In multiphase porous media, the heat flux is given by:

$$\mathbf{q}_{T} = -K_{e}\nabla T + n\sum_{\gamma}^{phase}\left(\rho^{\gamma}C_{p}^{\gamma}\right)T\mathbf{v} \tag{3.73}$$

where $K_e$ is the thermal conductivity, $T$ is the temperature, $C$ is the specific heat capacity of the fluid phase, and $\mathbf{v}^{\gamma}$ is the velocity of the fluid phase.

According to the definition of heat flux, the governing equation for heat transfer is:

$$\sum_{\gamma}^{phase}\left(\rho^{\gamma}C_{p}^{\gamma}\right)\frac{\partial T}{\partial t} - \nabla \mathbf{q}_{T} = Q_{T} \tag{3.74}$$

Boundary conditions are as follows:

$$\mathbf{q}_{T}\cdot n = \mathbf{q}_{T}|_{\Gamma},\ T = T_{\Gamma},\ \forall x \in \partial\Omega \tag{3.75}$$

Initial conditions are:

$$T(x) = T_{0}(x),\ \forall x \in \partial\Omega \tag{3.76}$$

In the equation, $Q_T$ denotes the heat source.

### 3.3.3 Auxiliary Process

The main and auxiliary processes include:

(1) The conventional definition of capillary pressure is the difference between the pressure fractions of the non-wetting and wetting fluids, as expressed by the following equation:

$$p^{\varsigma} = p^{CO_2} - p^l \tag{3.77}$$

(2) $S_{eff}$ is defined as the effective wetting fluid saturation:

$$S_{eff} = \frac{S^l - S^l_{res}}{1 - S^l_{res} - S^{CO_2}_{res}} \tag{3.78}$$

where $S^l_{res}$ denotes the residual wetting phase saturation, and $S^{CO_2}_{res}$ denotes the residual non-wetting phase saturation.

(3) The relationship between relative permeability and saturation is as follows:

$$k^l_{res} = \left(S_{eff}\right)^{(2+3\lambda)/\lambda} \tag{3.79}$$

$$k^{CO_2}_{res} = \left(1 - S_{eff}\right)^2\left(1 - \left(S_{eff}\right)^{(2+\lambda)/\lambda}\right) \tag{3.80}$$

In addition to presenting the fundamental structural relationships characterising dielectric properties, appropriate state equations for fluid properties related to pressure (such as density and viscosity) are also considered.

(4) The strain tensor is related to the displacement vector through the elastic relationship of infinitesimal deformation, and is defined as:

$$\varepsilon^{\xi}\left(u^{\xi}\right) = \frac{1}{2}\left(\nabla u^{\xi} + \left(u^{\xi}\right)^T\right) \tag{3.81}$$

(5) For the analysis of stress at a point in isotropic, homogeneous, and linear elastic solids, the effective stress in the solid can be expressed by the generalised Hooke's law as follows:

$$\sigma^s_E = 2\mu^s\varepsilon^s + \lambda^s\left(sp\varepsilon^s\right)I \tag{3.82}$$

(6) Among them, $\mu^s$ and $\lambda^s$ are the Lamé constants of the elastic material, which are determined experimentally as follows:

$$\mu^s = \frac{E}{2(1+v)} = G \tag{3.83}$$

$$\lambda^s = \frac{Ev}{(1+v)(1-2v)} = \frac{2Gv}{1-2v} \tag{3.84}$$

where $E$ is the Young's modulus of the material, $v$ is the Poisson's ratio of the material, and $G$ is the shear modulus.

### 3.3.4 Fracture Criterion Equation

Rocks, as porous media, are composed of a solid framework, interconnected pore spaces, and fluids (gases and water) stored within the pore spaces of the framework. Pore pressure refers to the pressure exerted or transmitted by fluids within the rock, while effective stress denotes the stress transmitted between solid particles in the rock. Numerous laboratory experiments have demonstrated that pore pressure and effective stress exert different influences on the deformation and failure of fluid-saturated or partially saturated porous solids.

In this paper, the anisotropic hydraulic properties of rocks are characterised by the relationships between porosity, mean effective stress, and the relationship between permeability and porosity. The relationship between porosity and mean effective stress is as follows:

$$\phi = (\phi_0 - \phi_r)\exp\left(5 \cdot 10^{-8} \cdot \sigma_M\right) + \phi_r \tag{3.85}$$

In the equation, $\phi_0$ represents the pore spacing under zero stress, and $\phi_r$ denotes the residual pore spacing under high stress. The average effective stress is defined by the principal stress as follows:

$$\sigma_M = \frac{1}{3}(\sigma_1 + \sigma_2 + \sigma_3) \tag{3.86}$$

$$\begin{cases} \sigma_1 = \sigma_1' + \alpha_b \overline{P} \\ \sigma_2 = \sigma_2' + \alpha_b \overline{P} \\ \sigma_3 = \sigma_3' + \alpha_b \overline{P} \end{cases} \tag{3.87}$$

The relationship between permeability and porosity is based on the following exponential function:

$$k = k_0 \exp[22.2(\varphi/\varphi_0 - 1)] \tag{3.88}$$

In the equation, $k_0$ represents the zero-stress permeability. The variation functions of porosity and permeability were obtained by comparison with the laboratory measurement data for sandstones provided by Davis.

The capillary pressure varies with the permeability and porosity induced by applied stress, in accordance with the Leverett function:

$$P_c = P_{c0}(S_l)\frac{\sqrt{k_0/\phi_0}}{\sqrt{k/\phi}} \tag{3.89}$$

Therefore, porosity, permeability, and capillary pressure all directly or indirectly depend on the average effective stress. By calculating the critical pressure that may lead to hydraulic fracturing or the critical pressure that may cause interlayer shear slip, it is possible to analyse the likelihood of caprock failure.

The underlying assumption is that once the fluid pressure exceeds the minimum principal compressive stress, hydraulic fracturing may occur. Therefore, the critical pressure for fracture initiation is:

$$P_{fc} = -\sigma_3 \tag{3.90}$$

Assuming that a slip can occur along a fracture plane in any arbitrary direction at any given point, the criterion for failure can be expressed in the following form:

$$|\tau_{m2}| = -(\sigma_{m2} + P_{sc})\sin\varphi + S_0\cos\varphi \tag{3.91}$$

$$\tau_{m2} = \frac{1}{2}(\sigma_1 - \sigma_3) \tag{3.92}$$

$$\sigma_{m2} = \frac{1}{2}(\sigma_1 + \sigma_3) \tag{3.93}$$

$S_0$ and $\varphi$ represent the internal cohesion coefficient and the internal friction angle of the fracture, respectively. Similar to the hydraulic fracturing criterion, the critical pressure for the initiation of shear fracture opening is given by:

$$P_{sc} = -\sigma_{m2} + S_0\cot\varphi - \frac{|\tau_{m2}|}{\sin\varphi} \tag{3.94}$$

An increase in pore pressure will raise both the horizontal effective stress and the vertical effective stress; however, this does not necessarily lead to an increase in the maximum in-plane shear stress, as it also depends on the constraint coefficient and the Poisson's ratio. Therefore, the horizontal Lamé criterion (3.94) and the vertical phase Lamé criterion (3.96) are introduced, both of which can be activated by an increase in pore pressure.

$$\sigma_h' \geq 0 \tag{3.95}$$

$$\sigma_v' \geq 0 \tag{3.96}$$

## 3.4 Section Summary

This chapter serves as a key link connecting the theoretical foundations and engineering practice of $CO_2$ geological storage, with a central focus on the development and application of numerical simulation techniques. It transforms the fundamental theories discussed in the previous two chapters—such as the properties of storage media, fluid displacement laws, and multiphysics coupling mechanisms—as well as the storage mechanisms in various scenarios, into quantifiable and predictable engineering tools. These tools are essential core supports for the practical engineering application of geological $CO_2$ storage.

Analyses have been conducted on the three major mainstream scenarios: $CO_2$-enhanced oil recovery ($CO_2$-EOR), enhanced coalbed methane recovery ($CO_2$-ECBM), and saline aquifer storage. For $CO_2$-EOR, a comprehensive control equation system was established by analysing the coupled mechanisms of seepage, stress, and damage, providing a precise model for predicting production enhancement, optimising reservoir parameters, and assessing storage security. For ECBM, the integrated relationships among adsorption, seepage, and temperature fields were considered, comprehensively elucidating the synergistic dynamic processes of $CO_2$ storage and methane recovery in coal seams. Regarding saline aquifer storage, scientific assumptions were employed to establish control and fracture determination equations, supporting large-scale storage, long-term safety assessment, and risk management.

## References


1. Li G (2016) Percolation–stress–damage coupling mechanism of $CO_2$ storage in depleted oil reservoirs and its impact on the safety of $CO_2$ storage. Southwest Petroleum University
2. Wang J (2014) Study on the elastoplastic damage MHC coupling model of sandstone and its numerical algorithm. Dalian Maritime University
3. Li J (2014) Study on the coupling of clay percolation–stress–damage and self-healing of permeability. Wuhan University
4. Tien C, Ramarao BV (2013) Can filter cake porosity be estimated based on the Kozeny-Carman equation? Powder Technol 237:233–240
5. Xie H, Peng J, Zeng Y et al (2005) Preliminary study on the energy distribution of sandstone failure. J Rock Mech Eng 15:2603–2608
6. Lu Y, Zhu L, Zheng Y (2000) Study on the characteristics of non-Darcy flow in ultra-low permeability reservoirs. Daqing Petrol Geol Dev 05:27–28+30–68
7. Liu H (2023) Numerical simulation study on geological storage of $CO_2$ in coalbed methane reservoirs in the southern demonstration project of the Shenmu Basin. China University of Mining and Technology
8. Fan L, Xu J (2020) Authority-enterprise equilibrium based mixed subsidy mechanism for carbon reduction and energy utilisation in the coalbed methane industry. Energy Policy 147:111828
9. Liu S, Fang H, Sang S et al (2020) $CO_2$ injectability and $CH_4$ recovery of the engineering test in Qinshui Basin, China based on numerical simulation. Int J Greenh Gas Control 95:102980

10. Guo H, Ni X, Wang Y et al (2018) Experimental study of $CO_2$-water-mineral interactions and their influence on the permeability of coking coal and implications for $CO_2$-ECBM. Minerals 8(3):117
11. Ma T, Rutqvist J, Oldenburg CM et al (2017) Coupled thermal-hydrological-mechanical modelling of $CO_2$-enhanced coalbed methane recovery. Int J Coal Geol 179:81–91
12. Fan J, Feng R, Wang J et al (2017) Laboratory investigation of coal deformation behaviour and its influence on permeability evolution during methane displacement by $CO_2$. Rock Mech Rock Eng 50(7):1725–1737
13. Fan Y, Deng C, Zhang X et al (2018) Numerical study of $CO_2$-enhanced coalbed methane recovery. Int J Greenh Gas Control 76:12–23
14. Liu X, Sang S, Zhou X et al (2023) Coupled adsorption-hydro-thermo-mechanical-chemical modelling for $CO_2$ sequestration and well production during $CO_2$-ECBM. Energy 262:125306
15. Mu Y (2018) Experimental study on the geochemical effects of $ScCO_2$ injection into coal seams and the structural response of the reservoir. China University of Mining and Technology, Xuzhou
16. Du Y, Fu C, Pan Z et al (2020) Geochemistry effects of supercritical $CO_2$ and $H_2O$ on the mesopore and macropore structures of high-rank coal from the Qinshui Basin, China. Int J Coal Geol 223:103467
17. Charriere D, Pokryszka Z, Behra P (2010) Effect of pressure and temperature on diffusion of $CO_2$ and $CH_4$ into coal from the Lorraine basin (France). Int J Coal Geol 81(4):373–380
18. Fang H, Sang S, Liu S (2019) Establishment of dynamic permeability model of coal reservoir and its numerical simulation during the $CO_2$-ECBM process. J Petrol Sci Eng 179:885–898
19. Zhang H, Liu J, Elsworth D (2008) How sorption-induced matrix deformation affects gas flow in coal seams: a new FE model. Int J Rock Mech Min Sci 45(8):1226–1236
20. Mu Y, Fan Y, Wang J et al (2019) Numerical study on the injection of heated $CO_2$ to enhance $CH_4$ recovery in water-bearing coal reservoirs. Energy Sources Part A Recov Util Environ Eff 1–19
21. Bi J (2023) Numerical simulation study on $CO_2$ storage in aquifers of the North Yellow Sea Basin. Dalian University of Technology
22. Rutqvist J, Börgesson L, Chijimatsu M et al (2001) Thermohydromechanics of partially saturated geological media: governing equations and formulation of four finite element models. Int J Rock Mech Min Sci 38(1):105–127

# Chapter 4
# Contributions of Geophysical Technologies to Carbon Dioxide Geological Storage

The global geophysical industry attaches great importance to its role in low-carbon development. While $CO_2$ geological storage presents opportunities for geophysical exploration, it also brings challenges. Geophysical technology is one of the key methods for investigating subsurface geological structures and stratigraphic properties, and can play a proactive and distinctive role in geological carbon storage projects. Its functions are mainly reflected in the following three aspects:

(1) Selection and Evaluation of Geological Storage Sites for Carbon Dioxide. As previously mentioned, effective geological storage of $CO_2$ requires favourable geological conditions, including suitable storage reservoirs, caprocks, and associated lithological and hydrogeological characteristics. During the preliminary assessment and planning and design stages of geological carbon storage projects, it is necessary to conduct a comprehensive characterisation of potential storage sites. This includes the evaluation of storage capacity, injectivity, and containment performance. The assessment of storage capacity primarily involves factors such as the volume, porosity, and spatial distribution of the storage formation (reservoir). Injectivity evaluation mainly concerns the permeability of the reservoir and the prevailing geostatic pressure conditions. The assessment of containment performance focuses on the permeability barrier properties of the caprock, its structural integrity and continuity, the sealing capacity of faults, the mechanisms of $CO_2$ retention, and capillary pressure, among other factors. Therefore, geophysical techniques are required to provide reliable information on subsurface geological structures, the petrophysical properties (porosity and permeability) of reservoirs and caprocks, and subsurface fluid data, thereby supplying dependable information and parameters for the evaluation of geological conditions and storage capacity of potential storage sites.

(2) Effectiveness monitoring and evaluation of geological storage of $CO_2$. After $CO_2$ is injected underground, it is necessary to employ geophysical and related technologies to dynamically monitor and assess the effectiveness of geological



D. Li and Y. Guo, *Geological Storage and Monitoring of Carbon Dioxide*,
https://doi.org/10.1007/978-981-92-3621-3_4

storage, in order to determine whether effective migration and trapping have occurred. By integrating geological, geophysical, and geochemical monitoring data with numerical simulation methods, it is possible to dynamically monitor and evaluate the migration patterns, migration directions, and spatial distribution of the injected $CO_2$ plume. This provides a basis for reference and decision-making in the formulation of subsequent $CO_2$ injection schemes. The main objectives of evaluating the effectiveness of geological storage are: to track the distribution range of $CO_2$ migration and the movement of subsurface fluids, to verify whether the $CO_2$ is retained within the target reservoir, to identify the mechanisms of $CO_2$ trapping and storage, to monitor changes in the storage space status, and to assess the residual storage capacity of $CO_2$.

(3) Safety monitoring and assessment of $CO_2$ geological storage. The structural stability of $CO_2$ geological storage sites and the potential for $CO_2$ leakage are the two main aspects of safety monitoring, specifically including: whether the injection of $CO_2$ for geological storage significantly alters the pressure conditions of subsurface strata, thereby compromising the stability of caprock and underground geological structures; whether it leads to volumetric expansion of strata resulting in surface deformation; whether it causes fracturing of the caprock; whether new faults are generated or existing faults are reactivated, affecting the integrity of the seal; whether it induces seismic activity; whether there is upward migration and diffusion of $CO_2$ into overlying strata, thereby causing disruption to shallow subsurface environments; and whether there is leakage of $CO_2$ into the atmosphere, among other concerns.

Geology, geophysics, geochemistry, and numerical modelling are the principal monitoring techniques for $CO_2$ geological storage. Long-term monitoring in geophysics and geochemistry can provide model parameters for numerical simulations. Geophysical monitoring methods are indirect approaches. As model-based monitoring techniques, they utilise geophysical monitoring data to assess whether the geophysical responses following $CO_2$ geological storage align with expectations, to infer the distribution of $CO_2$ plumes and their impact on subsurface conditions from monitoring data, or to detect and evaluate the risk of $CO_2$ leakage and induced seismicity. At different stages of geological carbon storage projects, there are significant differences in application scenarios, conditions, objectives, and technical requirements, necessitating the use of corresponding geophysical methods and technologies.

## 4.1 Application of Geophysical Techniques in the Selection and Evaluation of Geological Carbon Storage Sites

During the preliminary demonstration and planning and design stages of geological carbon storage projects, it is necessary to select and evaluate potential geological storage sites for $CO_2$. The primary tasks of geophysics are to provide technical support for the following geological tasks: (1) Imaging and evaluation of subsurface geological structures, mainly including the assessment of whether features such as anticlines, fault blocks, fractures, lithological traps, and stratigraphic traps can form effective carbon storage traps; (2) Evaluation of the porosity, permeability (injectivity), and connectivity of the target reservoir; (3) Identification and evaluation of reservoir fluids in the target formation to assess the mechanisms of carbon storage; (4) Evaluation of the effectiveness of the caprock overlying the target reservoir; (5) Assessment of the storage capacity (storage potential) of the target reservoir; (6) Evaluation of the subsurface stress state and geological structural stability to assess the risk of secondary disasters resulting from $CO_2$ injection. The suitability of the reservoir and caprock lithology, as well as structural stability, are key criteria for selecting geological carbon storage sites. Therefore, lithological testing has become an important task in the preliminary demonstration of geological storage projects. In the selection and evaluation of geological carbon storage sites, the application conditions and objectives are similar to those for reservoir description in oil and gas exploration. As a result, commonly used geophysical reservoir description techniques can generally be applied without significant modification.

Before the implementation of geological carbon storage projects, feasibility studies must be conducted to assess the long-term effectiveness and safety of geological storage through simulation and modelling platforms. Specific predictive indicators should be established, and effective and safe monitoring systems should be planned and designed accordingly. This is not only a requirement for obtaining approval from regulatory authorities, but also necessary for ensuring transparency to the public and gaining public understanding.

Modelling is an essential tool for demonstrating the feasibility of geological carbon storage projects. Feasibility studies of geological carbon storage require the establishment of three sets of modelling systems: first, numerical simulation of fluid dynamics for the migration and storage processes following underground injection of $CO_2$; second, numerical simulation of subsurface state changes (such as stress) after $CO_2$ injection; and third, forward modelling of geophysical responses (including seismic, electromagnetic, and gravity) following the injection, migration, and storage of $CO_2$. Based on rock physics theory, the integrated workflow from geomechanical numerical simulation of the reservoir to forward modelling of seismic responses forms a unified process for feasibility analysis and dynamic monitoring and evaluation of geological carbon storage.

Strictly speaking, geological storage of $CO_2$ is a thermodynamic-hydrodynamic-mechanical-chemical (THMC) coupled response process within a porous medium. This represents a complex modelling challenge involving multi-scale temporal and spatial dimensions, as well as the combined effects of multiple physical systems. Developing such an integrated coupled response simulator is a complex and highly challenging task, and remains an unfinished endeavour; current simulations have been significantly simplified.

The exploration and assessment of underground carbon storage spaces involve multiple technologies:

(1) Identification and Interpretation of Geological Characteristics

This category of technology encompasses the interpretation of information such as faults, strata, and lithological characteristics from seismic data, well-logging data, gravity data, electrical data, high-resolution aeromagnetic data, and remote sensing data. Among these, seismic data and well-logging data are the most widely used technical methods in various demonstration projects and research studies, as they provide a wealth of information. These methods not only enable the identification of the location and distribution of strata and faults, but also allow for the analysis of parameters related to lithology, sedimentary environment, heterogeneity, and fault activity. This facilitates the creation of key maps reflecting structural and sedimentary features, such as formation thickness maps, structural contour maps of various horizons, and sedimentary facies maps. Commonly used software includes LandMark and Geoframe for seismic interpretation, as well as Resform and Shiweng for well-logging interpretation.

(2) Indoor Testing and Experiments

Based on various research and demonstration cases, this type of technology primarily encompasses the following categories: testing of mineral compositions in geological strata and rock masses; analysis of groundwater and formation water compositions; measurement of reservoir and caprock parameters, with commonly tested parameters including porosity, permeability, degree of consolidation, pore size, and grain size; experiments investigating the development of fractures and deformation in geological strata and rock masses under different stress conditions; and experiments on $CO_2$–water–rock interactions. These tests and experiments are all conducted using core samples obtained from actual drilling, providing fundamental parameters for feasibility, suitability, and safety assessments, as well as for numerical modelling, and facilitating the analysis of sealing mechanisms.

(3) Three-Dimensional Modelling and Numerical Simulation

The approach to three-dimensional modelling of geological bodies for $CO_2$ storage is similar to that used for oil and gas reservoirs. Both are based on data such as well locations, well trajectories, stratification, seismic interpretation profiles, and geological characteristics (including thickness, sand content, porosity, permeability, and contour maps of various parameters). Typically, the process begins with the construction of a structural model (such as formation tops and bases, and faults) using deterministic modelling methods (various interpolation techniques). This is followed by the application of stochastic modelling methods (such as fractal, sequential Gaussian, sequential indicator, and truncated Gaussian methods), combined with well log data to establish facies models (sedimentary facies and lithofacies). On the basis of the facies model, similar methods are then used to construct property models (porosity, permeability, displacement pressure, etc.). The significance of these three modelling steps for $CO_2$ geological storage is not identical: the structural model determines the stratigraphic framework and three-dimensional skeletal characteristics of the geological storage body; the facies model not only effectively indicates regions within the geological body where sand bodies are well developed, but also reveals the vertical and lateral configuration relationships between reservoir and caprock; the property model, in addition to quantitatively characterising reservoir heterogeneity, also identifies favourable storage zones with good physical properties. The choice of interpolation methods in deterministic modelling and the application of various simulation techniques in stochastic modelling should be determined according to specific geological characteristics and actual operational results, as there are no absolute standards.

## 4.2 Application of Geophysical Techniques in the Effectiveness Monitoring and Evaluation of Geological Carbon Storage Projects

During the implementation phase of geological carbon storage projects, it is necessary to conduct effective monitoring and evaluation of geological storage of $CO_2$. The primary tasks of geophysics are to provide technical support for the following geological objectives: (1) monitoring the migration patterns, directions, and spatial distribution of $CO_2$ plumes; (2) identifying the mechanisms of $CO_2$ capture and storage; (3) monitoring changes in the state of the $CO_2$ storage space (reservoir); and (4) assessing the residual storage capacity of $CO_2$. However, current applications are mainly limited to the first aspect.

The effectiveness and safety monitoring of $CO_2$ geological storage are characterised by their dynamic nature. Dynamic monitoring requirements dictate that geophysical measurements must involve long-term repeated observations or continuous monitoring. Time-lapse geophysical methods, such as time-lapse seismic, time-lapse electromagnetic, and time-lapse gravity techniques, have become the principal approaches. The monitoring spatial coverage includes surface monitoring, cross-well monitoring, and borehole monitoring.

### 4.2.1 *Time-lapsed Seismic Monitoring*

$CO_2$ injection induces changes in the elastic parameters of geological formations, providing a theoretical basis for seismic monitoring methods. By analysing the temporal and spatial variations of the seismic wavefield, it is possible to indirectly monitor $CO_2$ migration and ascertain the spatial distribution of $CO_2$. Depending on the type of seismic source and observation method, seismic monitoring typically includes surface seismic, borehole seismic, and microseismic techniques.

(1) Ground-based seismic monitoring

Surface seismic monitoring utilises elastic waves generated by seismic sources at the surface, which propagate downwards and are reflected at impedance interfaces. The velocity and waveform of these waves are influenced by the impedance characteristics of the rock-fluid system, resulting in variations. Time-lapse seismic monitoring involves repeated seismic surveys at the same observation points but at different times, enabling the effective monitoring of fluid fronts and fluid dynamic characteristics. This, in turn, allows for the verification of whether $CO_2$ is effectively sealed within the target reservoir and for the detection of any potential $CO_2$ leakage or migration.

Figure 4.1 presents the time-lapse seismic profiles and time slices at the Sleipner site, where geological storage experiments of $CO_2$ have been conducted. Since 1996, $CO_2$ has been injected at a nearly constant rate into an underground saline aquifer approximately 1000 m below sea level. By 2011, around $1.3 \times 10^4$ tonnes of $CO_2$ had been stored. Baseline monitoring was carried out in 1994 prior to injection, with repeat surveys conducted in 1999, 2001, 2004, 2006, 2008, and 2010. The seismic sections clearly reveal the spatial distribution and dynamic evolution of the $CO_2$ plume. At this experimental site, $CO_2$ is injected into unconsolidated sandstone containing saline water, resulting in a decrease in P-wave velocity by up to 60%, thereby producing a pronounced time-lapse seismic anomaly. At the Weyburn oil field in Canada, enhanced oil recovery is achieved by injecting $CO_2$ and water, and time-lapse seismic imaging can effectively delineate the distribution of $CO_2$ within the carbonate reservoir. Figure 4.2 shows the relative $CO_2$ saturation distribution map derived from time-lapse seismic data.

Numerical simulation of reservoir fluid flow, petrophysical modelling, and seismic wavefield numerical simulation are essential tools for the analysis of $CO_2$ geological storage monitoring data. These methods enable the assessment of whether seismic

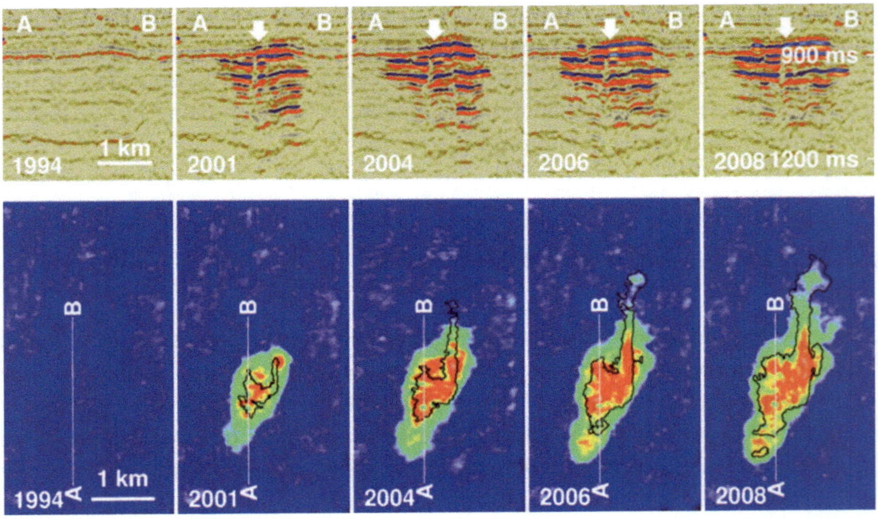


**Fig. 4.1** Time-lapse seismic profile and horizontal time slices for the Sleipner site

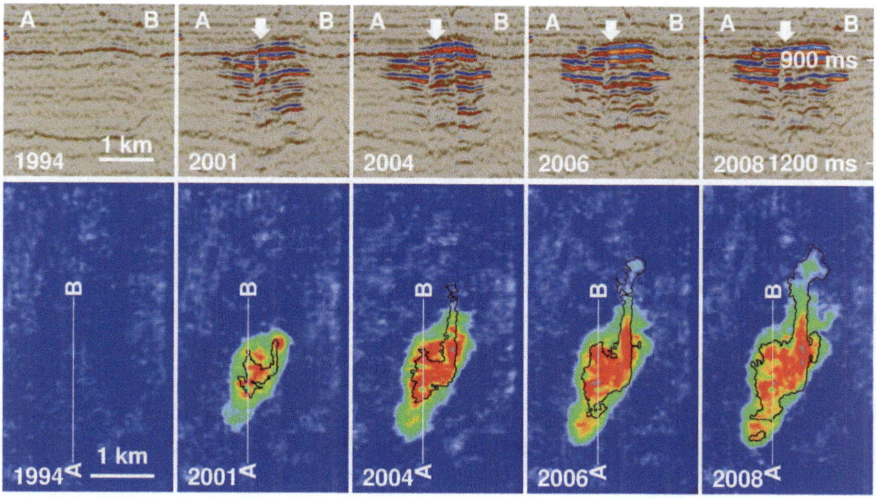


**Fig. 4.2** Plan-view prediction of relative $CO_2$ saturation from time-lapse seismic data in the Weyburn oilfield

monitoring can effectively detect the plume distribution resulting from $CO_2$ injection. Analysis based on petrophysical modelling from the Otway Project in Australia indicates that, when evaluating the sensitivity of fluid variations to petrophysical properties, it is necessary to fully consider the impact of heterogeneity in petrophysical characteristics. In feasibility analyses of time-lapse seismic monitoring, it is important not only to account for seismic anomalies caused by changes in reservoir properties, but also to consider the noise level in seismic observations. The presence of seismic observation noise can render certain time-lapse seismic anomalies in some operational areas indistinct.

Time-lapse seismic monitoring is one of the most important methods for geological carbon storage. It can be used to monitor the depth and location of injected $CO_2$, the spatial distribution of $CO_2$ migration and dispersion, as well as potential leakage at faults and caprock interfaces. A key factor affecting the effectiveness of time-lapse seismic applications is the repeatability of multiple seismic surveys. High-precision, permanently installed seismic data acquisition equipment, together with corresponding data processing workflows, are crucial for improving the repeatability of time-lapse seismic surveys and for obtaining high signal-to-noise ratio, high-resolution time-lapse seismic anomalies.

At present, the effectiveness of geological carbon storage monitored by time-lapse seismic surveys is primarily demonstrated through repeated seismic observations of amplitude anomalies (i.e., variations relative to baseline measurements). Other approaches include using three-dimensional or four-dimensional seismic data to invert for changes in $CO_2$ saturation or equivalent thickness distribution; identifying anomalies through AVO (amplitude versus offset) attribute analysis; and predicting fracture development orientation via seismic amplitude azimuthal anisotropy analysis. In some fields, anomalies in reflection travel time caused by decreases in seismic velocity have also been observed (see Fig. 4.3). After $CO_2$ injection into the storage formation, various petrophysical parameters of the reservoir undergo changes, which are typically reflected in seismic attributes. Seismic inversion can estimate changes in reservoir petrophysical parameters based on variations in seismic attributes. Research indicates that, in multiphase flow reservoirs formed by $CO_2$ injection, porosity and $CO_2$ saturation have a strong influence on the seismic wave frequency dispersion in sandstones, while temperature and pressure have a significant impact on seismic wave velocity. Based on multiphase flow Biot theory, $CO_2$ saturation distribution can be derived from seismic monitoring data inversion. Similarly, seismic data inversion can yield parameters such as clay content, porosity, static Young's modulus, and static Poisson's ratio. These reservoir parameters can be used for constructing geomechanical and fluid flow models of the reservoir and for numerical simulations, thereby enabling the characterisation and evaluation of the reservoir, caprock, and overburden during the $CO_2$ injection process. However, the inversion of $CO_2$ distribution or reservoir parameter changes from seismic monitoring data for storage remains subject to numerous challenges.

(2) Downhole Seismic Monitoring

Borehole seismic technology follows the same principles as surface seismic methods, but the detectors or seismic sources are placed within the borehole. Surface measurements are taken at a considerable distance from the subsurface geological reservoirs, resulting in weaker signals and greater surface interference. In contrast, borehole measurements offer a significantly improved signal-to-noise ratio in geophysical observations compared to surface measurements. However, the high cost of drilling limits the number, density, and placement of boreholes, making it difficult to deploy large-scale and widely distributed geophysical monitoring systems.

Based on differences in observation methods, downhole seismic monitoring includes Vertical Seismic Profiling (VSP) and cross-well seismic surveys. VSP is

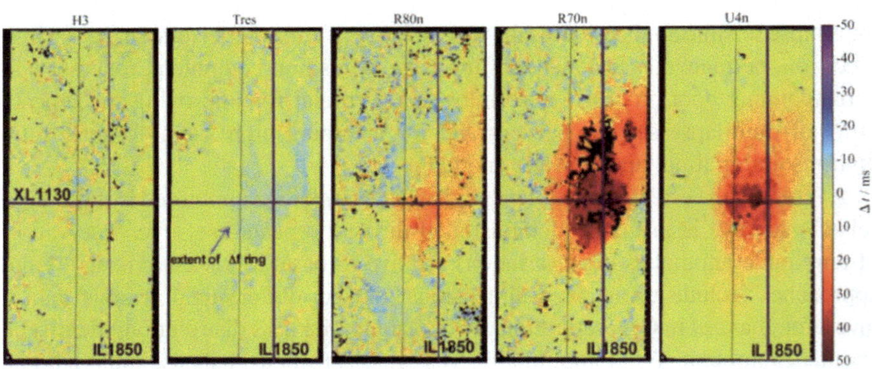


**Fig. 4.3** Reflection travel-time anomalies at different horizons in the Sleipner site

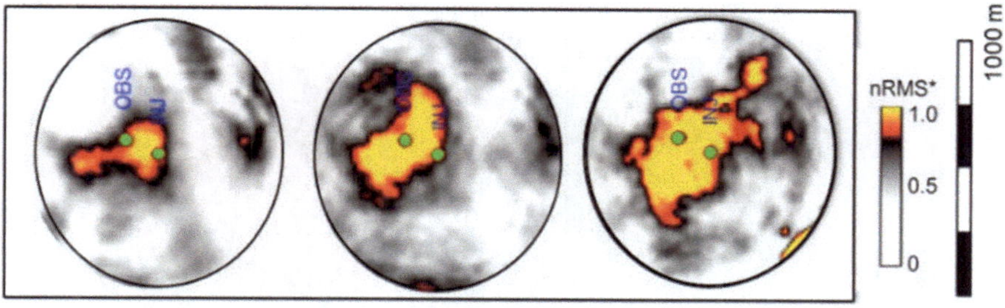


**Fig. 4.4** Monitored extent of the $CO_2$ plume after injection of 36,000, 102,000, and 141,000 t of $CO_2$

a seismic exploration technique in which geophones are placed within the borehole for data acquisition. This approach reduces the attenuation of high-frequency seismic signals caused by the low-velocity near-surface layer, thereby providing higher resolution compared to surface seismic surveys. During the geological storage of $CO_2$, VSP can provide critical information on subsurface geological structures, as well as the concentration and migration of $CO_2$ in the vicinity of individual wells. This technique plays a vital role in reservoir characterisation and time-lapse monitoring, making it the preferred method for tracking $CO_2$ migration. Significant progress has been achieved in projects such as the Frio project in Texas, USA, the Otway project in Australia, the Weyburn and Aquistore projects in Canada, and the Shenhua CCS demonstration project in China. At the Aquistore project, since 2013, multiple baseline and monitoring datasets have been acquired using VSP, which have been applied to $CO_2$ reservoir imaging, tracking the extent of the $CO_2$ plume, and verifying the integrity of the caprock. As shown in Fig. 4.4, from left to right, the monitored plume extents correspond to $CO_2$ injection volumes of 36,000, 102,000, and 141,000 tonnes, respectively. The results demonstrate that VSP monitoring methods can clearly delineate the boundaries of $CO_2$ plume dispersion.

Inter-well seismic monitoring involves placing the seismic source and receivers in adjacent wells to conduct monitoring, enabling the acquisition of information on inter-well reservoir structure, fluid distribution, injection effectiveness, and trace $CO_2$ leakage, thereby facilitating the monitoring of $CO_2$ migration and distribution between wells. This technique requires the injection well and monitoring wells to be relatively close to each other. To ensure high-resolution monitoring, the arrangement of multiple monitoring wells must be considered simultaneously, resulting in relatively high costs. Both the Nagaoka and Frio projects employed inter-well seismic techniques to accurately obtain information on the subsurface migration of $CO_2$, with significant results. The outcomes of inter-well seismic monitoring at the Frio site are shown in Fig. 4.5. Based on the results of inter-well seismic observations, the vertical profile of P-wave velocity changes between the injection and monitoring wells was derived (Fig. 4.5a), and the inter-well profile of $CO_2$ saturation was estimated using a fluid substitution model (Fig. 4.5b).

(3) Microseismic Monitoring

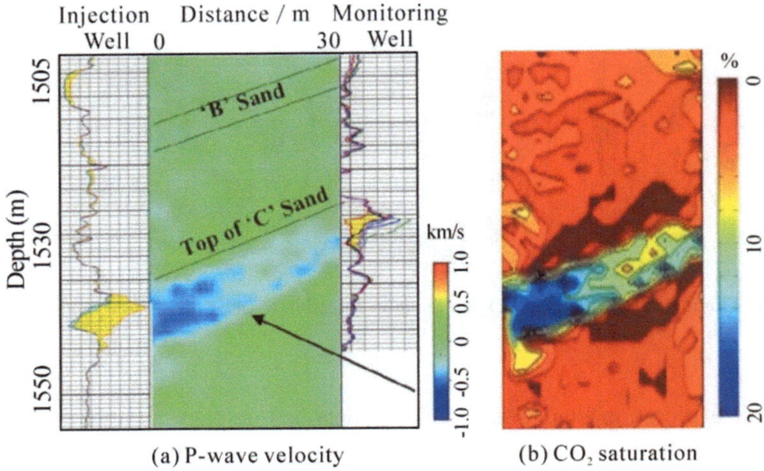


**Fig. 4.5** Crosswell seismic P-wave velocity change profile and $CO_2$ saturation profile at the Frio site

Groundwater extraction, geothermal resource utilisation, underground mining, and oil and gas exploitation are all human activities that may induce earthquakes. The assessment of seismic risk associated with anthropogenic activities, as well as the monitoring and early warning of induced seismicity, have become significant challenges in the field of geoscience, with some countries even introducing mandatory requirements in this regard. In the context of $CO_2$ geological storage monitoring, microseismic technology—widely applied in hydraulic fracturing monitoring during shale oil and gas development—currently offers a unique method for providing real-time, three-dimensional subsurface information. Passive-source microseismic monitoring can detect microseismic events generated at the leading edge of $CO_2$ plume migration prior to injection, and can also be used to monitor earthquakes induced by fault activation resulting from $CO_2$ injection. Microseismic monitoring can be conducted at the surface, within boreholes, or through integrated surface and borehole arrays. By means of microseismic monitoring, the locations of microseismic activity can be determined, thereby enabling estimation of the pressure front generated by fluid injection as it propagates through subsurface strata.

Passive-source microseismic monitoring in the Weyburn field, Canada, indicates a clear correlation between microseismic events and the rate of $CO_2$ injection. The spatial distribution of microseismic events also shows a good correlation with the distribution of the delayed seismic monitoring of the $CO_2$ plume migration. However, overall, the seismic events are not intense, suggesting that $CO_2$ injection has not caused significant geomechanical deformation of the reservoir. In addition, microseismic monitoring has detected shear-wave splitting phenomena, which can be used to estimate the orientation of fractures. Figure 4.6 shows the planar distribution of microseismic events in the Weyburn field from August 2003 to January 2006.

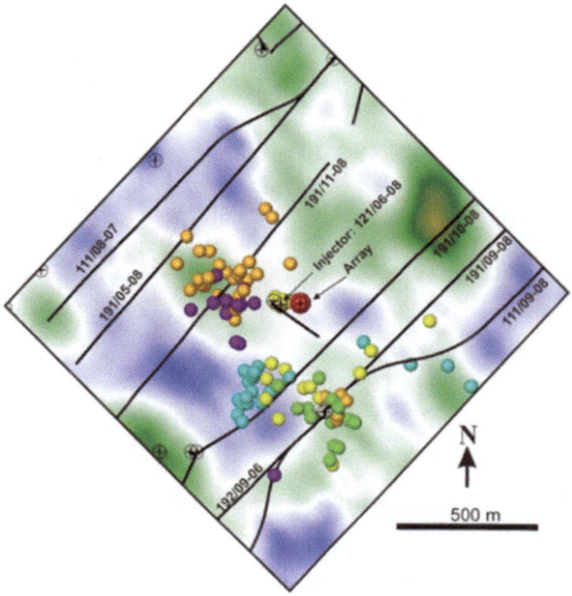


**Fig. 4.6** Plan-view distribution of microseismic events at the Weyburn site

Ambient noise seismic interferometry is an effective passive-source seismic data analysis method. Based on microseismic monitoring data from different periods, it is possible to obtain the time-lagged seismic response of the reservoir during the $CO_2$ storage process. Numerical simulation experiments and actual observation results from the Ketzin site in Germany have verified this approach, but it requires that the distribution of ambient noise sources possesses good illumination characteristics. Ambient noise seismic interferometry has been applied in seismic monitoring experiments at the Aquistore site in Canada; however, satisfactory results have not been achieved, possibly due to insufficient body wave energy and limited signal illumination directionality.

### *4.2.2 Time-lapsed Gravity Monitoring*

Gravity monitoring involves tracking the temporal and spatial variations in the gravitational acceleration field at $CO_2$ geological storage sites, thereby indirectly enabling the monitoring of subsurface $CO_2$ plume migration and revealing the spatial distribution patterns of the underground $CO_2$ plume. Following the injection of $CO_2$ into the subsurface reservoir, the injected $CO_2$ displaces the original pore fluids within the reservoir. If there is a density difference between the two, the density of the reservoir before and after $CO_2$ injection will change, resulting in corresponding alterations in the gravitational acceleration field. If this density difference is sufficient to be resolved by gravity monitoring, then the detection of gravity anomalies can be used

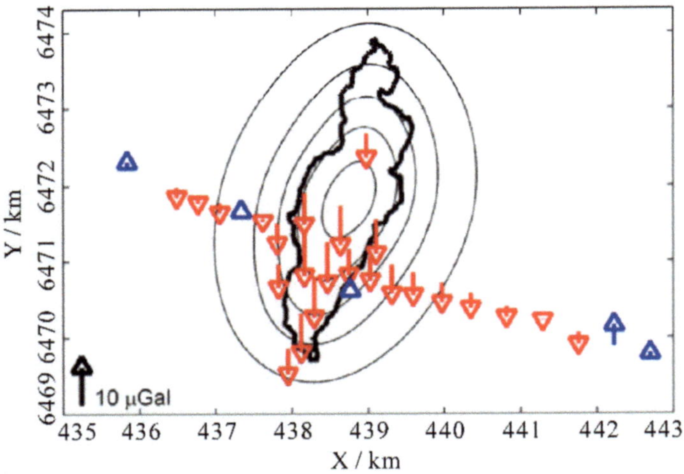


**Fig. 4.7** Observed gravity changes at the Sleipner site

to monitor $CO_2$ migration. The time-lapse gravity method has been successfully applied at the Sleipner site, where nearly two decades of seabed gravity observations have yielded clear time-lapse gravity responses to the $CO_2$ plume, demonstrating that, under current instrument precision, highly repeatable and reliable observational results can be obtained. As shown in Fig. 4.7, estimates of $CO_2$ density based on gravity measurements indicate that the annual dissolution rate of $CO_2$ in brine is less than 1.8%, confirming the effectiveness of gravity measurements for $CO_2$ monitoring.

The unique advantage of gravimetric monitoring over other monitoring techniques lies in its ability to directly estimate the mass changes and distribution of $CO_2$ following injection, requiring no additional conversions except for the application of the Newtonian gravitational constant. Consequently, such data can be directly utilised for quantitative estimation and flow model calibration. Gravimetric monitoring can therefore be regarded as a technique that complements well log time series (pressure and flow rate) as well as seismic or electromagnetic methods.

### 4.2.3 *Time-lapsed Electromagnetic Monitoring*

Electrical and electromagnetic (EM) techniques are used to measure variations in electric and magnetic fields induced by electric currents. Electrical methods, also known as resistivity techniques, utilise only current and voltage measurements, with frequencies low enough that EM induction effects can be neglected. In contrast, electromagnetic techniques require frequency-dependent sources to induce currents underground. Electromagnetic monitoring methods achieve indirect monitoring of the spatial and temporal variations of the electromagnetic field in $CO_2$ geological storage sites, thereby enabling the detection of subsurface $CO_2$ spatial distribution patterns. These methods are particularly suitable for monitoring $CO_2$ geological storage in deep saline aquifers. Electromagnetic monitoring approaches include electrical resistivity tomography (ERT), electromagnetic imaging, and controlled-source electromagnetic (CSEM) methods, among others.

ERT employs advanced computer control technology, relying on powerful data processing capabilities and digital imaging techniques. It integrates multi-device and multi-electrode spacing combined measurement methods that unify electrical sounding and electrical profiling. In $CO_2$ geological storage monitoring, ERT utilises the changes in reservoir fluid resistivity following $CO_2$ injection to achieve subsurface $CO_2$ migration imaging. At the Ketzin site in Germany, surface-to-borehole electrical resistivity tomography (SD-ERT) was used to monitor $CO_2$ injection and migration. Electromagnetic measurement data from the surface to the subsurface were obtained via one injection well and two observation wells. As shown in Fig. 4.8, the time-lapse inter-well electrical resistivity tomography (ERT) results indicate that the resistivity in the injection zone increased significantly by 200% compared to the background value. It is estimated that the injection zone is located at a depth of approximately 635 m, with an overall $CO_2$ saturation of 50% [35]. In the vertical array, the data resolution corresponds to the number of electrodes at a spacing of 10 m; therefore, small-scale fingering patterns and low $CO_2$ saturation zones were not detected.

Electromagnetic monitoring technology can provide information on the subsurface resistivity distribution, which can be interpreted to estimate the distribution of $CO_2$ saturation. The data resolution of ERT technology depends on the spacing between electrodes; inter-well ERT is more sensitive to resistivity changes near the wells, while surface-to-borehole ERT and CSEM measurements can extend the lateral coverage and provide data for tracking $CO_2$ plume migration. As the dependence of resistivity on $CO_2$ saturation differs from that of seismic velocity, electrical and electromagnetic techniques serve as a complement to seismic monitoring.

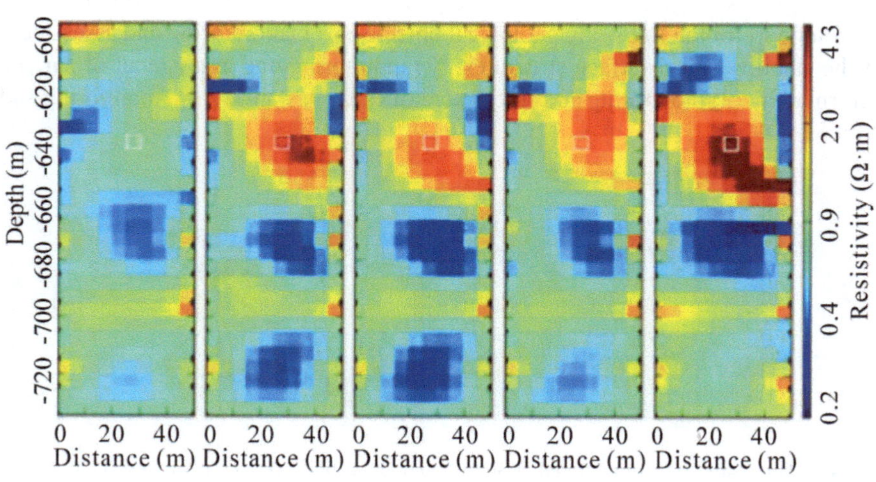


**Fig. 4.8** Time-lapse resistivity inversion results at the Ketzin site

### 4.2.4 Time-lapsed Distributed Acoustic Sensing Monitoring

Distributed Acoustic Sensing (DAS) is a novel technology that utilises fibre optic sensing to detect vibrations. In recent years, DAS technology has developed rapidly and has become an economically efficient solution particularly well-suited for permanent installation and long-term monitoring. It is applicable to both active-source and passive-source seismic observations. In geological carbon storage projects, DAS can be employed for time-lapse monitoring and imaging at various stages of $CO_2$ injection. It can be used to monitor the migration plume of injected $CO_2$, as well as to detect fault slip and induced seismicity triggered by the injection process. Passive-source seismic monitoring data can also be used to invert subsurface velocity models. Moreover, DAS is especially suitable for long-term monitoring when permanently deployed in wells or beneath the surface. The permanent installation of fibre optic sensors in wells can enhance the repeatability of time-lapse seismic measurements and allows for the simultaneous acquisition of DAS and Distributed Temperature Sensing (DTS) data. Figure 4.9 presents the results of a two-dimensional time-lapse DAS VSP survey in the Quest field, where the amplitude anomalies clearly delineate the outline of the injected $CO_2$ plume.

## 4.3 Application of Geophysical Techniques in Safety Monitoring and Evaluation of Geological Carbon Storage Projects

During the implementation and subsequent stages of geological carbon storage projects, it is necessary to conduct long-term monitoring and assessment of the safety of geological storage of $CO_2$. The primary tasks of geophysics are to provide technical support for the following geological objectives: (1) monitoring and assessment of the stability of subsurface structures; (2) monitoring and assessment of surface deformation; (3) monitoring and assessment of caprock stability; (4) monitoring and

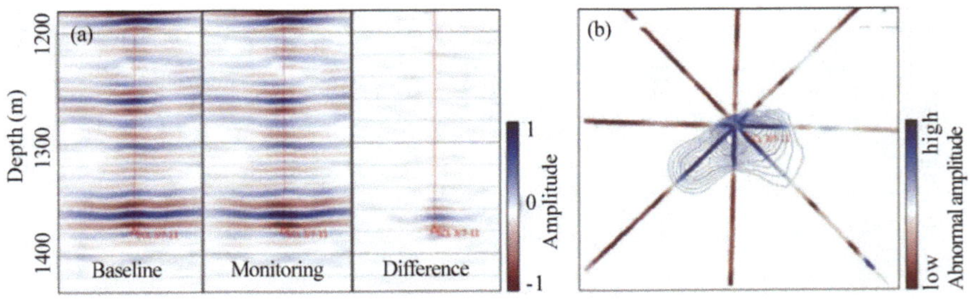


**Fig. 4.9** Time-lapse DAS-monitored VSP amplitude profile (**a**) and plan-view contours of amplitude anomalies (**b**) at the Quest site

assessment of fault stability; (5) monitoring of induced seismicity; and (6) monitoring of $CO_2$ leakage.

Current geophysical methods available for safety monitoring include three-dimensional seismic, microseismic, near-surface electromagnetic, and satellite remote sensing monitoring technologies. At present, safety monitoring mainly targets surface deformation, $CO_2$ leakage, and induced seismicity, while significant technical challenges remain in the monitoring of subsurface structural stability. Due to the limited number of practical cases of geophysical monitoring in geological carbon storage projects, there are few documented instances of actual safety incidents.

### *4.3.1 Surface Deformation Monitoring*

In recent years, space technology has developed rapidly. Satellite-based Earth observation has become a low-cost, large-scale, and high-frequency technique for monitoring surface environmental changes. Geodetic methods can detect surface deformation anomalies induced by changes in subsurface reservoir pressure. Global Navigation Satellite Systems (GNSS) enable three-dimensional spatial positioning and displacement estimation at the millimetre scale on the Earth's surface, which can be used for displacement monitoring at surface control points. Interferometric Synthetic Aperture Radar (InSAR), based on satellite observation, utilises the phase delay of reflected microwaves or radar waves to estimate the displacement at surface observation points, achieving millimetre-level measurement accuracy. InSAR can perform area-based observations in a scanning mode, allowing for low-cost monitoring and analysis of surface deformation anomalies. This, in turn, facilitates the assessment of the impact of geological carbon storage on subsurface structural stability and the surface environment, serving as a low-cost, long-term monitoring tool for the geological safety of carbon storage.

Interferometric synthetic aperture radar (InSAR) has yielded promising results in several geological carbon storage projects. For example, its application in the Algyo and Salah enhanced gas recovery demonstration projects has revealed the migration of injected $CO_2$ along faults and fracture zones, with observable surface deformation detected above the injection wells (Fig. 4.10). All three injection wells exhibited surface uplift, with an average annual increase of 3 mm. Notably, above the KB-502 injection well, seismic reflection data also indicated anomalous travel-time pull-up in narrow, banded zones, corresponding to the migration of injected $CO_2$ along faults and fracture zones.

Further research can establish a relational model between surface deformation and subsurface reservoir volume changes induced by fluid injection. Based on observed surface deformation data, it is possible to inversely estimate the spatial distribution of fluid volume changes within the reservoir.

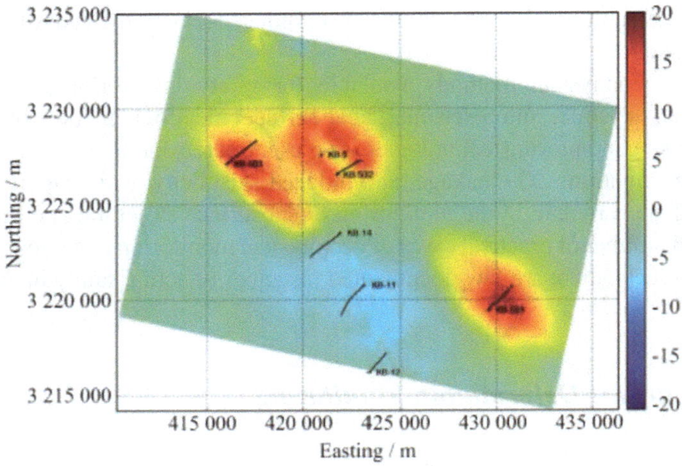


**Fig. 4.10** Surface vertical deformation obtained from InSAR observations for the Salah project

## *4.3.2 Carbon Dioxide Leakage Monitoring*

Surface $CO_2$ emission measurements can be used to assess natural emissions under different geological conditions, as well as to monitor industrial pollution emissions or leakage in $CO_2$ geological storage projects. A portable laser remote sensing method can obtain one-dimensional $CO_2$ concentration profiles through angular scanning, and then use multiple one-dimensional profiles acquired from multi-point scanning to perform tomographic inversion, thereby obtaining a two-dimensional planar concentration distribution. The resulting data represent the surface area $CO_2$ emission intensity, which can be used to locate $CO_2$ emission or leakage sources. Ground-based multi-spectral observation data from satellites and unmanned aerial vehicles can also be used to monitor surface emissions and leakage of greenhouse gases such as $CO_2$ and methane. Figure 4.11 shows the planar distribution map of $CO_2$ emission concentration obtained by tomographic inversion, based on one-dimensional $CO_2$ concentration profiles measured by a scanning laser remote sensing instrument at three measurement points at the Solfatara volcano in Italy.

Surface soil resistivity measurements and near-surface resistivity tomography results can serve as a tool for detecting surface $CO_2$ leakage points. The distribution of near-surface soil resistivity exhibits a clear correlation with both the concentration and flux of $CO_2$.

## *4.3.3 Microseismic Event Monitoring*

After $CO_2$ is injected into the storage formation, under the combined effects of the geostatic pressure gradient and buoyancy, the $CO_2$ plume undergoes both lateral and

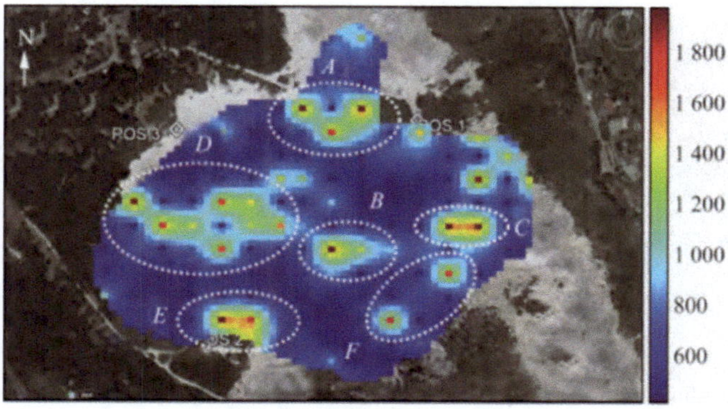


**Fig. 4.11** Plan-view distribution of $CO_2$ emission concentration obtained by remote-sensing tomography inversion

vertical migration, resulting in pressure disturbances and, in some cases, localised pressure accumulation within the strata. This can induce the opening or formation of fractures in the storage and overlying caprock, and may even activate faults, leading to microseismic events of varying magnitudes. By employing microseismic monitoring techniques to detect such events within the strata, it is possible to assess the safety of $CO_2$ containment. Microseismic monitoring at the Weyburn oilfield has identified multiple microseismic events of magnitude 1–3 occurring in both the overlying and underlying strata of the storage formation. In the Illinois Basin saline aquifer $CO_2$ storage project, 10,123 microseismic events of magnitude 1–2 were detected in the lower strata of the storage formation during a 22-month injection period. Analysis suggests that the injected $CO_2$ activated several minor faults in the area (see Fig. 4.12). In ultra-low permeability oil reservoirs, the high injection pressure of $CO_2$ and the significant pressure disturbances generated during migration increase the risk of fracture activation. Therefore, focused microseismic monitoring of the storage and caprock, combined with analysis of the underlying causes, is essential for evaluating the integrity of the caprock following $CO_2$ injection.

## 4.4 Section Summary

This chapter systematically describes the core supporting role of geophysical technologies throughout the entire lifecycle of $CO_2$ geological storage, focusing on site selection and evaluation, assessment of storage effectiveness, and safety monitoring. In the selection and evaluation of storage sites, technologies such as seismic exploration, well logging analysis, gravity and electromagnetic surveys are employed to accurately characterise the geological structure of potential reservoirs, the sealing capacity of caprocks, and storage capacity, thereby providing quantitative data to support the identification of safe and efficient storage sites. Laboratory experiments

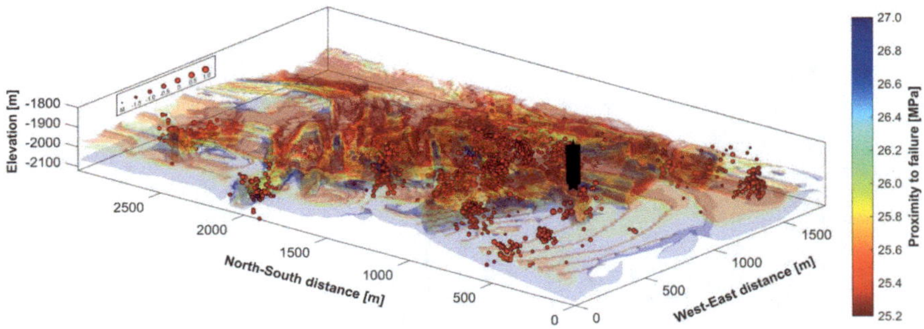


**Fig. 4.12** Three-dimensional display of microseismic events during injection; black indicates the injection well and red indicates microseismic events

and numerical simulation techniques are also introduced. For monitoring the effectiveness of $CO_2$ storage, time-lapse seismic monitoring can track the migration pathways and distribution range of $CO_2$; time-lapse gravity monitoring can invert the cumulative mass of $CO_2$ within the reservoir; time-lapse electromagnetic monitoring can identify the extent of plume dispersion and changes in saturation; and time-lapse distributed acoustic sensing enables real-time, high-precision dynamic capture of the injection process. The integration of multiple technologies ensures that the storage process is both observable and controllable. In terms of safety assurance for $CO_2$ storage, this chapter also introduces three key monitoring technologies: surface deformation monitoring, which uses InSAR or ground-based measurements to detect surface changes and provide early warning of abnormal pressure; $CO_2$ leakage monitoring, which utilises gas detection and electromagnetic imaging to identify leakage pathways; and microseismic event monitoring, which can assess the fracturing characteristics of the storage body and caprock.

Overall, geophysical technologies underpin the entire carbon storage engineering chain, establishing a technical safeguard system from early-stage decision-making to long-term operation. This serves as a key support for promoting large-scale application of carbon storage and facilitating the achievement of the "dual carbon" targets, providing an important technological pathway for global climate governance.

## References

1. Zhao G (2023) Carbon dioxide geological storage geophysical monitoring: current status, challenges and future developments. Pet Geol Exp 62(2):194–211
2. Singleton S (2021) Leading the way to a carbon-neutral world. Lead Edge 40(10):712–713
3. SEG (2021) SEG position on climate change. Lead Edge 40(3):222–223
4. Ghosh R, Sen MK (2012) Predicting subsurface $CO_2$ movement: from laboratory to field scale. Geophysics 77(3):M27–M37
5. Zakharova NV, Goldberg DS, Olsen PE et al (2020) Reservoir and sealing properties of the Newark rift basin formations: implications for carbon sequestration. Lead Edge 39(1):38–46

6. JafarGandomi A, Curtis A (2011) Detectability of petrophysical properties of subsurface $CO_2$-saturated aquifer reservoirs using surface geophysical methods. Lead Edge 30(10):1112–1121
7. Meng C, Fehler M (2021) The role of geomechanical modelling in the measurement and understanding of geophysical data collected during carbon sequestration. Lead Edge 40(6):413–417
8. Meadows M (2008) Time-lapse seismic modelling and inversion of $CO_2$ saturation for storage and enhanced oil recovery. Lead Edge 27(4):506–516
9. Mur A, Barajas-Olalde C, Adams DC et al (2020) Integrated simulation to seismic and seismic reservoir characterisation in a $CO_2$ EOR monitoring application. Lead Edge 39(9):668–678
10. Juhlin C, Giese R, Zinck-Jørgensen K et al (2007) 3D baseline seismics at Ketzin, Germany: the $CO_2$ SINK project. Geophysics 72(5):B121–B132
11. Li J, Liner C, Geng P et al (2013) Convolutional time-lapse seismic modelling for $CO_2$ sequestration at the Dickman oilfield, Ness County, Kansas. Geophysics 78(3):B147–B158
12. Hao Y, Yang D, Zhou Y (2016) A feasibility study of $CO_2$ geologic sequestration integrating reservoir simulation, rock-physics theory, and seismic modelling. Geophysics 81(5):M71–M82
13. Ourabah A, Chatenay A (2022) Unlocking ultra-high-density seismic for CCUS applications by combining nimble nodes and agile source technologies. Lead Edge 41(1):27–33
14. Williams G, Chadwick A (2012) Quantitative seismic analysis of a thin layer of $CO_2$ in the Sleipner injection plume. Geophysics 77(6):R245–R256
15. Roach LAN, White DJ, Roberts B (2015) Assessment of 4D seismic repeatability and $CO_2$ detection limits using a sparse permanent land array at the Aquistore $CO_2$ storage site. Geophysics 80(2):WA1–WA13
16. Pevzner R, Caspari E, Gurevich B et al (2015) Feasibility of $CO_2$ plume detection using 4D seismic: CO2CRC Otway Project case study—part 2: detectability analysis. Geophysics 80(4):B105–B114
17. Wang YC, Morozov IB (2020) Time-lapse acoustic impedance variations during $CO_2$ injection in Weyburn oilfield, Canada. Geophysics 85(1):M1–M13
18. Kazemeini SH, Yang C, Juhlin C et al (2010) Enhancing seismic data resolution using the prestack blueing technique: an example from the Ketzin $CO_2$ injection site, Germany. Geophysics 75(6):V101–V110
19. Fawad M, Rahman MDJ, Mondol NH (2021) Seismic-derived geomechanical properties of potential $CO_2$ storage reservoir and caprock in Smeaheia area, northern North Sea. Lead Edge 40(4):254–260
20. Duxbury A, White D, Samson C et al (2012) Fracture mapping using seismic amplitude variation with offset and azimuth analysis at the Weyburn $CO_2$ storage site. Geophysics 77(6):B295–B306
21. White D (2009) Monitoring $CO_2$ storage during EOR at the Weyburn-Midale Field. Lead Edge 28(7):838–842
22. Verdon JP, Kendall JM, White DJ (2010) Passive seismic monitoring of carbon dioxide storage at Weyburn. Lead Edge 29(2):200–206
23. Cheraghi S, White DJ, Draganov D et al (2017) Passive seismic reflection interferometry: a case study from the Aquistore $CO_2$ storage site, Saskatchewan, Canada. Geophysics 82(3):B79–B93
24. White D, Daley TM, Paulsson B et al (2021) Borehole seismic methods for geologic $CO_2$ storage monitoring. Lead Edge 40(6):434–441
25. Yang D, Malcolm A, Fehler M et al (2014) Time-lapse walk-away vertical seismic profile monitoring for $CO_2$ injection at the SACROC enhanced oil recovery field: a case study. Geophysics 79(2):B51–B61
26. Hu LW, Doetsch J, Brauchler R et al (2017) Characterising $CO_2$ plumes in deep saline formations: comparison and joint evaluation of time-lapse pressure and seismic tomography. Geophysics 82(4):ID1–ID18
27. Sidenko E, Tertyshnikov K, Gurevich B et al (2022) Distributed fibre-optic sensing transforms an abandoned well into a permanent geophysical monitoring array: a case study from Australian South West. Lead Edge 41(2):140–148

28. White D, Bellefleur G, Dodds K et al (2022) Toward improved distributed acoustic sensing sensitivity for surface-based reflection seismics: configuration tests at the Aquistore $CO_2$ storage site. Geophysics 87(2):P1–P14
29. Gasperikova E, Li YG (2021) Time-lapse electromagnetic and gravity methods in carbon storage monitoring. Lead Edge 40(6):442–446
30. Vasco DW, Dixon TH, Ferretti A et al (2020) Monitoring the fate of injected $CO_2$ using geodetic techniques. Lead Edge 39(1):29–37
31. Comola F, Janna C, Lovison A et al (2016) Efficient global optimisation of reservoir geomechanical parameters based on synthetic aperture radar-derived ground displacements. Geophysics 81(3):M23–M33
32. Mathieson A, Midgley J, Dodds K et al (2010) $CO_2$ sequestration monitoring and verification technologies applied at Krechba, Algeria. Lead Edge 29(2):216–222
33. Queiber M, Burton M, Granieri D (2018) Large-area quantification of subaerial $CO_2$ anomalies with portable laser remote sensing and 2D tomography. Lead Edge 37(3):222a1–222a9
34. Dubucq D, Turon L, Blanco B et al (2021) Earth observation remote sensing for oil and gas: a new era. Lead Edge 40(1):26–34
35. Bondarenko N, Podladchikov Y, Williams-Stroud S et al (2025) Stratigraphy-induced localisation of microseismicity during $CO_2$ injection in Illinois Basin. J Geophys Res Solid Earth 130:e2024JB029526

# Chapter 5
# Fine Characterization Technology for Carbon Dioxide Storage Bodies

The effective geological storage of $CO_2$ requires that the reservoir possesses good sealing conditions, with no leakage pathways, and that the reservoir pressure, porosity, permeability, and caprock stability are all within suitable ranges. Small-scale geological structures serve as potential pathways for $CO_2$ leakage and seepage; therefore, high-precision and high-efficiency prediction of geological structures is an effective means of evaluating storage sites. The elastic parameters of the reservoir are key to characterising its properties, making the accurate acquisition of these parameters a crucial objective in exploration. Pre-stack seismic data contain abundant information on amplitude variations, and utilising this information enables the effective determination of sensitive reservoir parameters. This section introduces key seismic exploration technologies for $CO_2$ geological storage, focusing on two aspects: intelligent identification of small-scale geological structures and high-resolution prediction of reservoir elastic parameters.

## 5.1 Intelligent Identification Technology for Geological Structures of Reservoirs

Geological structure identification is a task characterised by ambiguity and uncertainty, typically requiring subjective human judgement and consuming a considerable amount of time. $CO_2$ geological storage demands a detailed understanding of geological structures at various scales within the storage area to provide a basis for engineering design. To address this, the present study investigates intelligent methods for geological structure identification by analysing multiple machine learning algorithms, and establishes a nonlinear mapping relationship between sensitive seismic attributes and geological structures.



D. Li and Y. Guo, *Geological Storage and Monitoring of Carbon Dioxide*,
https://doi.org/10.1007/978-981-92-3621-3_5

### 5.1.1 Data Preparation and Selection of Seismic Attributes

Forward modelling is an effective means of obtaining the seismic reflection responses of known geological structures. In this study, forward modelling data were used to analyse the sensitivity of seismic attributes to the target geological structures. Firstly, a typical geological model was established (as shown in Fig. 5.1), which included three types of lithology: from top to bottom, these are mudstone, coal, mudstone, and siltstone. The coal seam has a thickness of approximately 10 m. The model incorporates three types of geological structures: collapse columns, normal faults, and reverse faults. The specific geological parameters are listed in Table 5.1. From left to right, the geological structures are collapse column X1, collapse column X2, normal fault F1, and reverse fault F2. The long-axis diameter of the coal seam at collapse column X1 is 20 m, while at collapse column X2 it is 45 m. The fault throw of fault F1 is 8 m.

A 60 Hz Ricker wavelet was used as the seismic source, and data were recorded using a self-excitation and self-receiving mode with a channel spacing of 1 m. To account for the impact of noise in real data, 5% Gaussian white noise was added to the synthetic data. The resulting reflected wave time section is shown in Fig. 5.2. It can be observed that the coal seam reflection is clearly identifiable on the time section. At the subsidence column, the amplitude of the coal seam reflection is attenuated, while at the normal fault, the variation in the coal seam reflected wave is relatively

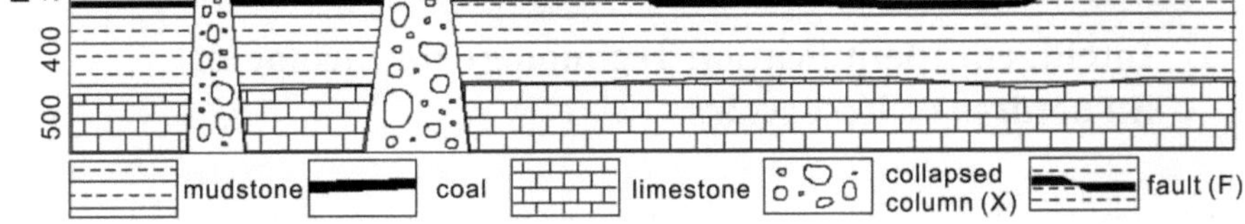


**Fig. 5.1** Geological model containing typical geological structures

**Table 5.1** Parameters of the geological model

| Parameter | Formation | | | Structure |
|---|---|---|---|---|
| | Mudstone | Coal | Limestone | Collapse column |
| Velocity (m/s) | 3000 | 2500 | 3600 | 1800 |
| Density (kg $m^{-3}$) | 2200 | 1500 | 2700 | 1100 |

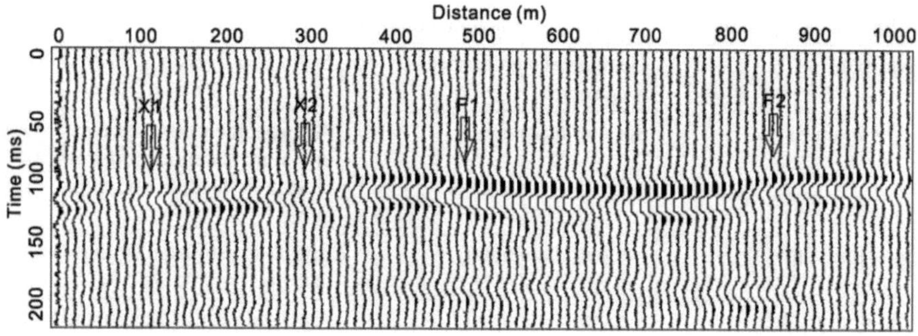


**Fig. 5.2** Stacked forward-simulated seismic section

minor. At the reverse fault, the coal seam reflected wave exhibits bending along the axis.

There are numerous types of seismic attributes, including amplitude, frequency, attenuation, and geometry, among others. Different seismic attributes carry distinct geological significance and therefore have varied applications, such as lithology identification, structural prediction, and fluid differentiation analysis. For geological structure identification tasks, in order to improve prediction accuracy and reduce data redundancy, it is necessary to optimise the selection of multiple seismic attributes.

In this book, the selection of seismic parameters is divided into the following three stages:

(1) Seismic attribute optimisation based on attribute algorithm theory. By integrating seismic datasets obtained from forward modelling, a comprehensive analysis of various seismic attributes was conducted. Thirteen theoretically feasible seismic attributes, which can potentially reflect structural discontinuities, were initially extracted (Table 5.2). These attributes are: 3D curvature, structural dip angle, dip illumination, first-order derivative, second-order derivative, variance, amplitude difference, chaos attribute, phase residual, dominant frequency, instantaneous frequency, instantaneous amplitude, and flatness attribute.
(2) Seismic attribute selection based on target geological structures. Building upon the thirteen seismic attributes initially extracted as described above, model data analysis was employed to examine the response characteristics of each attribute at known geological structures (faults and collapse columns). As illustrated in Fig. 5.3, six seismic attributes extracted along the coal seam are presented. It can be observed that the structural dip attribute exhibits relatively minor differences between structured and unstructured locations, whereas other attributes display either significantly high or low values at structural positions. Based on these considerations, eight seismic attributes were selected, including: 3D curvature, dip illumination, variance, chaos, dominant frequency, instantaneous frequency, instantaneous amplitude, and average energy attribute.

**Table 5.2** Seismic attributes sensitive to structural continuity

| Attribute | Geological significance represented | Definition |
|---|---|---|
| 3D curvature | Structural continuity | At a specific point on a curve, curvature is defined as the rate of change in the direction of the curve |
| Structural dip | Structural continuity | Local dip estimated from seismic data |
| Dip illumination | Structural continuity | A new dip-estimation method that displays the computed results in two different views |
| First derivative | Structural continuity | First derivative of the input seismic data volume |
| Second derivative | Structural continuity | Second derivative of the input seismic data volume |
| Variance volume | Structural continuity, lithology | Locally estimated variance in the signal |
| Amplitude contrast | Structural continuity, bed thickness, lithology, and fluid differences | This attribute is based on a Sobel filter and attempts to extract and highlight structures in the seismic data |
| Chaos volume | Structural continuity and fluid differences | A measure of disorganization in dip and azimuth estimation methods |
| Cosine of phase | Structural continuity | Cosine of the instantaneous phase, also referred to as normalized amplitude |
| Dominant frequency | Structural continuity, bed thickness, lithology, and fluid differences | Geometric mean of instantaneous frequency and instantaneous bandwidth |
| Instantaneous frequency | Structural continuity, bed thickness, lithology, and fluid differences | Calculated from the temporal rate of change of instantaneous phase |
| Instantaneous quality | Structural continuity, bed thickness, lithology, and fluid differences | Instantaneous quality is the ratio of instantaneous frequency to twice the instantaneous bandwidth |
| Flatness | Structural continuity, lithology | Variation in the directional field, used to determine signal uniformity within the range of direction estimation |

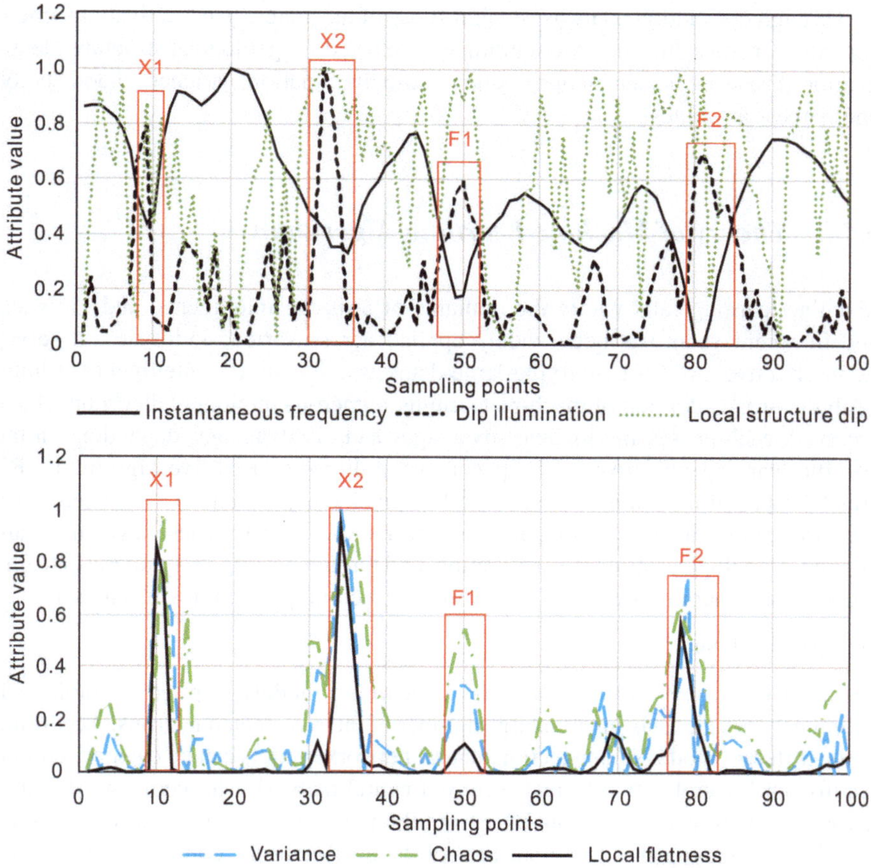


**Fig. 5.3** Comparison of responses of different attributes at geological structures

(3) Analysis of seismic attribute independence based on Q-type clustering. The clustering analysis method is applied to classify and optimise the eight selected attributes mentioned above. Firstly, the seismic attributes are normalised, then the distance coefficients between each attribute are calculated according to Eq. (5.1), and finally, the classification and merging are carried out.

$$d(x, y) = \sqrt{\sum_{k=1}^{p} (x_k - y_k)^2} \quad (5.1)$$

In the equation, $d$ denotes the distance coefficient, $p$ represents the number of properties, and $x$ and $y$ refer to the two properties used for calculating the correlation.

Through correlation analysis of eight types of attributes, five relatively independent and sensitive attributes were ultimately selected for geological structure identification. These are instantaneous frequency, dip illumination, variance, chaotic body, and flatness attributes.

### *5.1.2 Theory of Machine Learning Algorithms*

Machine learning falls within the domains of artificial intelligence and statistics, and in recent years has been widely applied across various industries. It serves as an effective tool for classifying large datasets and enabling intelligent learning. With the broad adoption of machine learning, numerous intelligent algorithms have emerged, each possessing its own advantages and disadvantages depending on the specific learning objectives. This paper primarily introduces five algorithms: BP (Back Propagation) neural networks, RBF (Radial-Basis Function) neural networks, support vector machines, decision trees, and random forests. These algorithms are applied to geological structure exploration, and through testing on synthetic data, the recognition capabilities of each algorithm in structural prediction are analysed.

(1) Artificial Neural Networks

Artificial neural networks are a series of learning models inspired by biological neural networks, consisting of a large number of interconnected neurons. According to the different modes of connection, neural network models can be categorised into feedforward neural networks and recurrent neural networks. In feedforward neural networks, neurons receive input data from the previous layer and transmit it to the next layer, meaning that the input to each layer of neurons consists solely of the output from the preceding layer, with no feedback between layers. In contrast, in recurrent neural networks, each neuron can be interconnected with others, allowing for the phenomenon of input signals being transmitted in loops, and thus recurrent neural networks inherently include feedback mechanisms. In this study, both the BP neural network and the RBF neural network employed belong to the category of feedforward neural networks, among which the BP neural network is a type of multilayer feedforward neural network that utilises error backpropagation.

The feedforward neural network comprises an input layer, a hidden layer, and an output layer, as illustrated in Fig. 5.4. After receiving input from the previous layer, each neuron computes the weighted sum of the input data and adds a bias term as the input value. In order to achieve linear classification of low-dimensional data, the neuron applies a nonlinear transformation to the input data via an activation function to obtain the output value. The calculation formula is as follows:

$$Z_j^l = \sum_{i=1\dots n} w_{ij} \cdot a_{ij}^{l-1} - b_j^l \tag{5.2}$$

$$a_j^l = f\left(z_j^{(l)}\right) \tag{5.3}$$

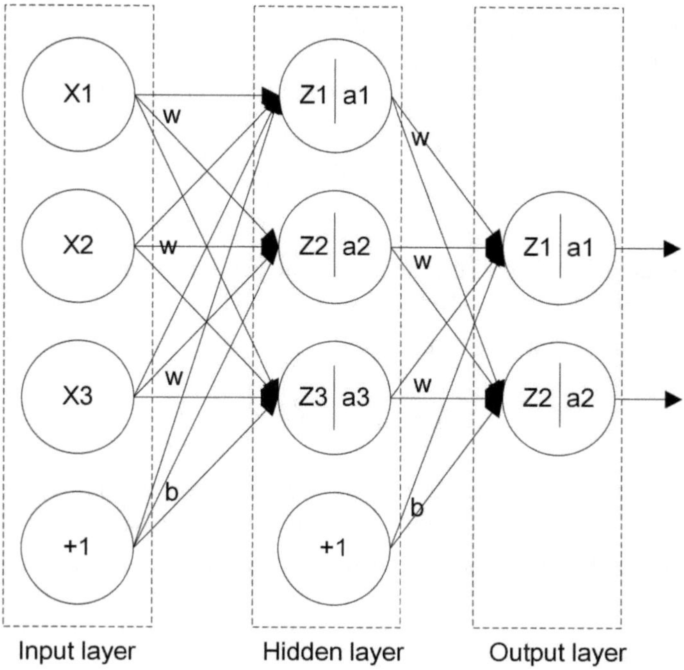


**Fig. 5.4** Neural network architecture

In the equation, $Z_j^l$ denotes the input value of the $l$th neuron in the $j$th layer, $w_{ij}$ represents the weight from the $i$th neuron to the $j$th neuron, al − 1 indicates the output of the l − 1 th layer, which also serves as the input to the $a^{l-1}$ th layer, $l-1$ denotes the bias (i.e., threshold), and $l$ represents the activation function.

After inputting the data, the BP neural network initially assigns weights and thresholds to each transmission line. It then performs forward propagation to obtain predicted values. If the prediction error exceeds the threshold, backpropagation is executed to update the weights and thresholds, followed by another round of forward propagation. This process is repeated until the error meets the specified requirement (set as error less than 0.01).

The computational procedure of the BP neural network is as follows:

(a) Calculation of the number of hidden layers based on the input and output layers:

$$l_2 = \sqrt{l_1 + l_3} + l_0 \tag{5.4}$$

In the equation, $l_1$, $l_2$, and $l_3$ represent the number of layers in the input layer, hidden layer, and output layer, respectively, and $l_0$ is an arbitrary constant.

(b) Assign initial values to the weights $w_{ij}$ and thresholds $b_j^l$ of each layer, where $w_{ij}$ denotes the weight from the $i$th neuron to the $j$th neuron, and $b_j^l$ represents the threshold of the $l$th neuron in the $j$ [− 1, 1] are chosen as small non-zero constants within the range of [− 1, 1].

(c) Calculation of the neuron values in the hidden layer and output layer $z_j$:

$$z_j = f\left(\sum_{i=1}^{n} x_i w_{ij} + b_j\right) \tag{5.5}$$

In the equation, $x_i$ denotes the input to the $i$th neuron, $f$ represents the activation function, which is generally calculated using fx = 11 + e − x.

(d) Calculate and output the layer error $d$:

$$d_j = z_j(1 - z_j) \cdot \left(z_j^P - z_j\right) \tag{5.6}$$

In the equation, $d_j$ denotes the error of the $j$th neuron, $z_j$ represents the output value of the $j$th neuron, and $z_j^P$ is the label (expected output value) of the $j$th neuron.

(e) Update weights and thresholds, with the variation calculated as follows:

$$\Delta w_{ij} = \lambda z_j d_j \tag{5.7}$$

$$\Delta b_j = \lambda d_j \tag{5.8}$$

In the equation, λ denotes the learning rate.

Repeat steps (c)–(e) until the error meets the required threshold, thus completing the learning process of the BP neural network. From the above description, it can be seen that the main characteristics of the BP neural network include:

1) A multilayer neural network structure comprising an input layer, hidden layers, and an output layer.
2) The transmission of information includes both forward and backward directions, while the neurons between layers are connected unidirectionally.
3) The activation function of a neuron is generally a sigmoid function, as shown in Eq. (5.5).
4) By continuously updating the weights and thresholds, the training and learning process of the network is completed.

Although BP neural networks have found widespread application and have achieved a certain degree of predictive success, their convergence rate is relatively slow, which presents challenges when processing large-scale datasets. The RBF neural network can be regarded as a type of nonlinear mapping from the input layer space (low dimensionality) to the hidden layer space (high dimensionality), followed by a linear mapping from the hidden layer space to the output layer. By selecting an appropriate RBF function, it is possible to address classification problems involving the transition from low to high dimensionality, thereby providing strong classification capability and a relatively high learning rate.

The learning process of the RBF neural network is generally divided into two steps. The first step is an unsupervised learning process, in which the weights from the input layer to the hidden layer are obtained through training; the second step is supervised learning, where the weights from the hidden layer to the output layer are determined through training. The computational procedure for RBF network learning is as follows:

(a) Initialisation parameters of the network, including the connection weights from the hidden layer to the output layer $w_{jk}$, the centre parameters of the neurons in the hidden layer $c_{ij}$, and the width vectors $d_{ij}$:

$$c_{ij} = \min i + \frac{\max i - \min i}{2p} + (j-1)\frac{\max i - \min i}{p} \tag{5.9}$$

$$d_{ij} = d_f \sqrt{\frac{1}{q}\sum_{k=1}^{q}\left(x_i^k - c_{ij}\right)} \tag{5.10}$$

In the equation, i = 1, 2, ... , o; j = 1, 2, ... , p and k = 1, 2, ... , q represent the number of neurons in the input layer, hidden layer, and output layer, respectively. $i = 1, 2, \ldots, o\ j = 1, 2, \ldots, p.$

(b) Calculation of the root mean square error of the network:

$$RMS = \sqrt{\frac{\sum_{i=1}^{o}\sum_{k=1}^{q}\left(z_k^P - z_k\right)^2}{qo}} \tag{5.11}$$

In the equation, $z_k^p$ denotes the expected output value of the $k$th neuron.

(c) Updating of network parameters: Here, the gradient descent method is employed to optimise the parameters of the RBF network. The iterative calculation method is as follows:

$$w_{jk}(t) = w_{jk}(t-1) - \eta\frac{\partial E}{\partial w_{jk}(t-1)} + \alpha\left[w_{jk}(t-1) - w_{jk}(t-2)\right] \tag{5.12}$$

$$c_{ij}(t) = c_{ij}(t-1) - \eta\frac{\partial E}{\partial c_{ij}(t-1)} + \alpha\left[c_{ij}(t-1) - c_{ij}(t-2)\right] \tag{5.13}$$

$$d_{ij}(t) = d_{ij}(t-1) - \eta\frac{\partial E}{\partial d_{ij}(t-1)} + \alpha\left[d_{ij}(t-1) - d_{ij}(t-2)\right] \tag{5.14}$$

In the equation, $w_{jk}$ denotes the weight from the $j$th neuron in the hidden layer to the $k$th neuron in the output layer; $c_{ij}$ represents the centre parameter from the $i$th neuron in the input layer to the $j$th neuron in the hidden layer; $d_{ij}$ indicates the width corresponding to $c_{ij}$; $t$ denotes the $t$th iteration; $\eta$ represents the learning rate; and $E$ is the network evaluation function:

$$E = \frac{1}{2}\sum_{i=1}^{o}\sum_{k=1}^{q}\left(z_k - z_k^P\right) \quad (5.15)$$

Repeat steps (b)–(c) until the error meets the required threshold, thereby completing the learning and training of the network.

Both BP neural networks and RBF neural networks are classified as feedforward networks and can be used interchangeably. However, the computational methods of the two networks differ. Compared with BP neural networks, RBF networks offer the following advantages:

1) Rapid learning speed, enabling integrated computation for large-scale data;
2) The network topology can automatically adapt according to specific problems, exhibiting strong self-organisation and self-learning capabilities;
3) The number of iterations and training time are significantly reduced, while approaching strong classification performance.

(2) Support Vector Machine

Support Vector Machine (SVM) is a supervised learning model based on statistical theory. Unlike neural networks, SVM employs mathematical methods and optimisation techniques. SVM maps complex vectors in a low-dimensional space into a higher-dimensional space, where it constructs an optimal separating hyperplane with the maximum margin, thereby achieving data classification (Fig. 5.5). SVM exhibits strong generalisation capability and performs effectively with small sample datasets.

Given a training sample dataset $(x, y)$, where xili = 1, 2, …, n represents the input samples and yili = 1, 2, …, n represents the output samples, a nonlinear mapping function φ is employed to project the data from a low-dimensional space to a high-dimensional space. Subsequently, the optimal decision function is constructed in the high-dimensional space as shown in Eq. (5.16):

$$y = \mathbf{w}^{\mathrm{T}}x + b \quad (5.16)$$

In the equation, $b$ denotes the displacement, and $\mathbf{w}$ represents the weighting vector.

The hyperplane can be represented by the equation $\mathbf{w}^{\mathrm{T}}x+b = 0$. Therefore, when data points in a two-dimensional space are mapped to an $n$-dimensional space, the distance from a point to the hyperplane can be expressed as:

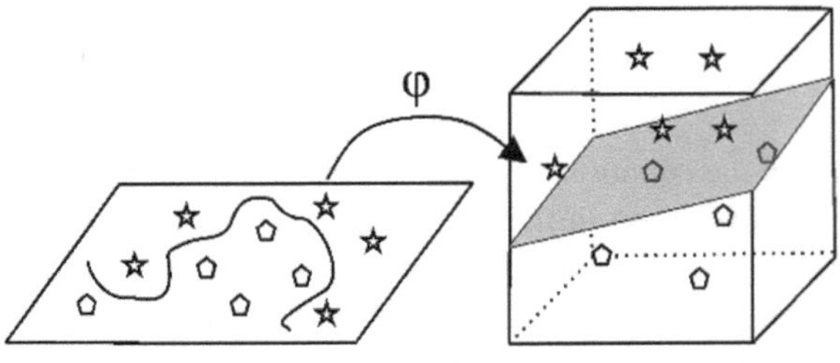


**Fig. 5.5** Construction of the hyperplane

$$\frac{\left|w^T x + b\right|}{\|w\|}, \|w\| = \sqrt{w_1^2 + w_2^2 + \cdots + w_n^2} \tag{5.17}$$

Assuming y = 1, − 1, the distance from the support vector to the hyperplane $y = 1, -1$ is given by the following relationship:

$$\begin{cases} \frac{w^T x + b}{\|w\|} \geq d, y = 1 \\ \frac{w^T x + b}{\|w\|} \leq -d, y = -1 \end{cases} \tag{5.18}$$

After transformation, the following can be obtained:

$$\begin{cases} \frac{w^T x + b}{\|w\| d} \geq 1, y = 1 \\ \frac{w^T x + b}{\|w\| d} \leq -1, y = -1 \end{cases} \tag{5.19}$$

Multiplying both sides of the equation by $y^2$ respectively, and combining the two expressions followed by transformation, yields:

$$d = \frac{y(w^T x + b)}{\|w\|} \tag{5.20}$$

Without affecting the outcome of the objective function, multiply the equation by 2 and take the maximum distance to obtain:

$$d = \max 2 * \frac{y\left(w^T x + b\right)}{\|w\|} \tag{5.21}$$

Since the support is oriented along $y(w^T x + b) = 1$, the distance formula becomes max 2‖w‖, i.e., min 12‖w‖. In order to facilitate taking the square root of ‖w‖, and without affecting the optimisation problem, it can be transformed into min 12‖w‖2.

The slack variable coefficient $C$ is used to represent the error between classes, while the penalty coefficient $\xi$ is employed to balance the margin and the error. At this point, the optimisation problem of the SVM algorithm can be expressed according to the minimisation principle as follows:

$$\begin{aligned} &\min \frac{1}{2}\|w\|^2 + C\sum_{i=1}^{n} \xi_i \\ &\text{s.t.} \quad \begin{cases} y_i\left(\mathrm{w}^{\mathrm{T}} x_i + b\right) \geq 1 - \xi_i \\ \xi_i \geq 0 \end{cases} \quad i = 1, 2, \ldots, m \end{aligned} \tag{5.22}$$

In the equation, $C$ denotes the penalty coefficient, $\|\cdot\|^2$ represents the relaxation coefficient, and $\xi$ indicates the L2 norm.

The above constrained optimisation problem can be solved using the Lagrange function:

$$L(x, \alpha, \beta) = f(x) + \sum_{i=1}^{k} \alpha_i c_i(x) + \sum_{j=1}^{m} \beta_j h_j(x) \tag{5.23}$$

In the equation, αi and $\alpha_i \beta_j$

$$L(\boldsymbol{w}, b, \alpha) = \frac{1}{2}\|\boldsymbol{w}\|^2 - \sum_{i=1}^{m} \alpha_i \left[ y_i(\boldsymbol{w}^T x_i + b) - 1 \right] \tag{5.24}$$

The deviation coefficient for both sides of the equation can be obtained as follows:

$$\begin{aligned} \nabla_w L(\boldsymbol{w}, b, \alpha) &= \boldsymbol{w} - \sum_{i=1}^{m} \alpha_i y_i x_i \\ \frac{\partial}{\partial b} L(\boldsymbol{w}, b, \alpha) &= \sum_{i=1}^{m} \alpha_i y_i \end{aligned} \tag{5.25}$$

Combining Eqs. (5.24) and (5.25) yields:

$$L(\boldsymbol{w}, b, \alpha) = \sum_{i=1}^{m} \alpha_i - \frac{1}{2} \sum_{i=1}^{m} \sum_{j=1}^{m} \alpha_i \alpha_j y_i y_j (x_i)^T x_j - b \sum_{i=1}^{m} \alpha_i y_i \tag{5.26}$$

$b \sum_{i=1}^{m} \alpha_i y_i = 0$ is evident that in Eq. (5.26), b∑i = 1mαiyi = 0.
According to Lagrange's theorem on parity, it follows that:

$$\begin{aligned} d^* &= \max \theta_D(\alpha, \beta) = \max \min L(\boldsymbol{w}, b, \alpha) \\ &= \min f(x) = \min \frac{1}{2}\|\boldsymbol{w}\|^2 \end{aligned} \tag{5.27}$$

Thus, the function with respect to $\alpha$ is obtained as follows:

$$\begin{aligned} &\max w(\alpha) = \sum_{i=1}^{m} \alpha_i - \frac{1}{2} \sum_{i=1}^{m} \sum_{j=1}^{m} \alpha_i \alpha_j y_i y_j < x_i, x_j > \\ &s.t. \sum_{i=1}^{m} \alpha_i y_i = 0, 0 \le \alpha_i \le C, i = 1, 2, \ldots, m \end{aligned} \tag{5.28}$$

In the equation, $< x_i, x_j >$ denotes the inner product of $(x_i)^T x_j$.
By solving Eq. (5.28), we obtain:

$$w = \sum_{i=1}^{m} \alpha_i y_i x_i$$
$$b = y_i - \sum_{i=1}^{m} y_i \alpha_i < x_i, x_j > \tag{5.29}$$

This solution process is designed for linear data calculations. When dealing with nonlinear data, in order to satisfy the mapping requirements, it is necessary to introduce a mapping function φx, whereby Eq. (5.16) is transformed into $\varphi(x)$. Assuming the kernel function is $y = w^T \cdot \varphi(x) + bk(x_i, x_j) = \varphi(x_i) \cdot \varphi(x_j)$.

The key aspect of SVM lies in the selection of the kernel function, as different kernel functions result in different SVM algorithms. Kernel functions can effectively address the complex computational problems associated with classification in high-dimensional spaces. Commonly used kernel functions include: linear kernel functions, sigmoid kernel functions, radial basis kernel functions, and polynomial kernel functions, among others.

(3) Decision Trees and Random Forests

Decision trees are a type of supervised learning method that utilise a tree-like structure, proceeding recursively from the top down. Starting at the root node of the tree, attribute values are tested and compared at internal nodes. Based on the attribute values of the given instance, the corresponding branches are determined, and ultimately, a conclusion is reached at the leaf nodes of the decision tree (see Fig. 5.6).

The general principle of constructing a decision tree is to calculate the information expectation of the sample dataset T = t1, t2, …, tm as follows:

$$E(T) = -\sum_{i=1}^{m} p(t_i) \log_2 p(t_i)$$
$$p(t_i) = \frac{|t_i|}{\sum_{i=1}^{m} |t_i|} \tag{5.30}$$

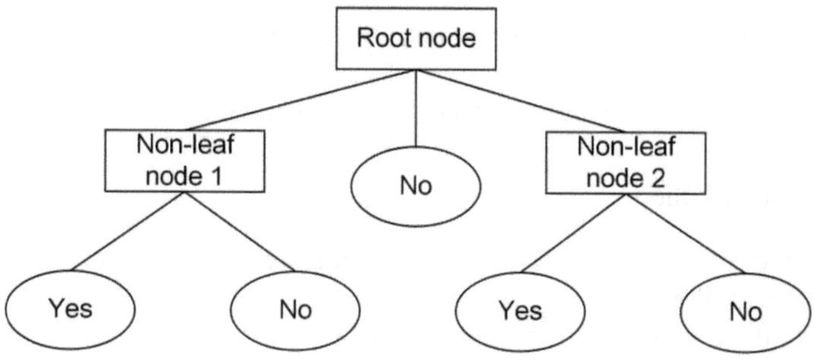


**Fig. 5.6** Structure of the decision-tree network

In the formula, $E$ denotes the expectation, $t$ represents the sample, and $m$ indicates the category. According to the probability distribution, $N$ independent samples are drawn to construct a classification tree such that the classification error rate $R_p$ is minimised. In a decision tree, both root and leaf nodes employ a feature condition $X_t$ to partition different classification outcomes. Therefore, the key to determining the effectiveness of a decision tree's classification lies in the selection of this optimal splitting feature. The current principal measures for evaluating the partitioning capability of such features mainly include the following three types:

(a) Information Gain

Information gain refers to the reduction in classification uncertainty based on the knowledge of a particular feature X. The main idea is to utilise the minimum possible amount of known information to obtain the maximum number of results. The calculation method for the information gain of each feature is as follows:

Assuming the current sample set is $E$, and the proportion of the $t$th class in the sample is $m$, then the information entropy at node $N$ can be expressed as follows:

$$H(D) = -\sum_{k=1}^{K} p_k \log_2(p_k) \tag{5.31}$$

When a discrete feature $A$ has $n$ possible values, the sample set $D$ can be partitioned according to these values, resulting in $n$ nodes. By independently calculating the weight and information entropy of each node, the empirical conditional entropy can be obtained as follows:

$$H(D|A) = \sum_{i=1}^{n} \frac{|D_i|}{|D|} H(D_i) = -\sum_{i=1}^{n} \frac{|D_i|}{|D|} \sum_{k=1}^{K} p_{ik} \log_2(p_{ik}) \tag{5.32}$$

Thus, based on the feature $A$, the information gain for partitioning the sample set $D$ at this node can be expressed as:

$$Gain(D, A) = H(D) - H(D|A) \tag{5.33}$$

Then the optimal partition criterion is expressed as:

$$A_* = \arg\max Gain(D, A), A \in (A_1, A_2, \ldots, A_m) \tag{5.34}$$

(b) Gini Index Method

The Gini index method is widely applied in feature selection, and its calculation procedure is as follows:

Calculate the Gini index of the sample set $D$:

$$Gini(D) = \sum_{k=1}^{K} p_k(1 - p_k) \tag{5.35}$$

In this equation, the meaning of $p_k$ is consistent with that in Eq. (5.31). Similarly, when the discrete feature $A$ has $n$ possible values, the Gini index of the sample set $D$ can be expressed as:

$$Gini\_index(D, A) = \sum_{i=1}^{n} \frac{|D_i|}{|D|} Gini(D_i) \tag{5.36}$$

At this point, the optimal partitioning criterion for the sample set is:

$$A_* = \arg\min Gini\_index(D, A), x \in X \tag{5.37}$$

(c) Incremental Benefit Ratio Method

The information gain method generally tends to favour features with a greater number of values. To address this shortcoming, improvements have been made to the information gain method, leading to the proposal of the gain ratio method, which can be expressed as follows:

$$\begin{aligned} Gain\_Ratio(D|A) &= \frac{Gain(D, A)}{H_A(D)} \\ H_A(D) &= -\sum_{i=1}^{n} \frac{|D_i|}{|D|} \log_2 \frac{|D_i|}{|D|} \end{aligned} \tag{5.38}$$

In the formula, $H_A(D)$ denotes the entropy of the sample set $D$ with respect to the feature $A$.

Although the decision tree algorithm is simple and offers fast classification, it still has certain shortcomings. For example, the repeated occurrence of subtrees can lead to redundancy in the decision tree. While the training error continues to decrease, the test error increases, resulting in overfitting. Therefore, in 2001, the random forest method was proposed. Random forest is an algorithm developed on the basis of decision trees and is a classifier that comprises multiple decision trees, effectively enhancing the generalisation capability of decision trees. The concept of ensemble learning addresses the limitations of single models by integrating multiple models, leveraging their strengths to compensate for weaknesses, thereby avoiding local limitations and improving classification accuracy.

By employing the Bagging strategy to perform resampling on the data, multiple datasets are generated, which are then used to construct multiple decision trees, thereby forming a random forest. This process involves two key random procedures: First, bootstrap sampling with replacement is applied to the original dataset to obtain multiple training and testing subsets, and an entirely independent decision tree is built for each training subset. Second, during the construction of each decision tree, the random forest method does not select the optimal feature for classification from all available features; instead, it randomly selects a subset of features and determines the best split among these. As random forests incorporate the ensemble concept based on the Bagging strategy, both the sample datasets and the features are randomly sampled, which helps to prevent overfitting.

The final classification decision formula of a random forest can be expressed as:

$$H(x) = \arg\max_{Y} \sum_{i=1}^{k} I(h_i(x) = Y) \tag{5.39}$$

In the equation, $H(x)$ denotes the ensemble classification model, $h_i(x)$ represents the classification model of the $i$th decision tree, $Y$ is the output variable, and $I(\cdot)$ refers to the indicator function.

### 5.1.3 Network Training

In order to generate datasets suitable for the prediction of faults and collapse columns, 1000 geological models were created using a randomised approach, as partially illustrated in Fig. 5.7. The random factors include the undulating morphology of geological strata, layer thickness, dip, as well as the location, size, and number of collapse columns and faults. Naturally, these random factors are constrained within a certain range to reflect actual geological conditions. For these 1000 models, seismic data were generated using a forward modelling approach with self-excitation and self-reception. After adding random noise of varying intensities, five types of sensitive seismic attributes were extracted. The input data for the network comprise the optimally selected five sensitive seismic attributes, while the label data represent the geological structures. For classification purposes, collapse columns are denoted by "– 5", faults by "5", and background data by "0", thus completing the training of the network.

### 5.1.4 Machine Learning Algorithm Optimisation

Using the trained network, predictions were made on the simulated data from Fig. 5.1 (which were not included in the training dataset). Along the coal seam, 100 sampling points were extracted, with each sampling point comprising five attributes. The prediction results of the five machine learning algorithms are shown in Fig. 5.8. As can be seen from the figure, all five machine learning algorithms achieved a certain level of predictive performance and demonstrated a reasonable degree of fit.

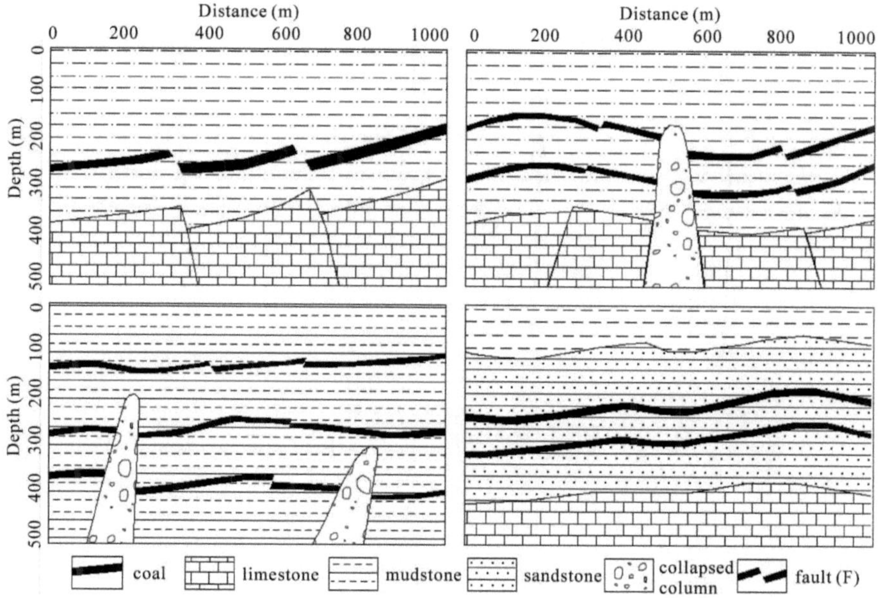


**Fig. 5.7** Schematic of randomly generated geological models

However, some errors still exist; for example, a fault was predicted near the subsidence column. This may be due to the occurrence of pull-down in the coal seam reflections adjacent to the subsidence column, resulting in features similar to those of a fault.

A comprehensive comparison of the advantages and disadvantages, prediction accuracy, and computational efficiency of each prediction method is presented (Table 5.3). This facilitates the selection of the most suitable algorithm for different geological tasks, research areas, and geological structures, thereby enabling efficient and high-precision seismic interpretation. In terms of prediction accuracy, the random forest method achieves the highest accuracy at 97%, while the radial basis function neural network (RBF) yields the lowest accuracy at 88%. Regarding computational efficiency, the decision tree demonstrates the highest efficiency, with a computation time of 2.7 s, whereas the support vector machine (GA-SVM) exhibits the lowest efficiency, with a computation time of 12 s. When generating each "tree", the random forest algorithm randomly selects only a subset of features, typically defaulting to $H(x)$ features ($h_i(x)$ being the total number of features), which ensures the randomness of the feature data and reduces overfitting. The random forest algorithm demonstrates strong anti-interference capability. Given the geological conditions of this study, the prediction results of the random forest algorithm outperform those of the other four algorithms.

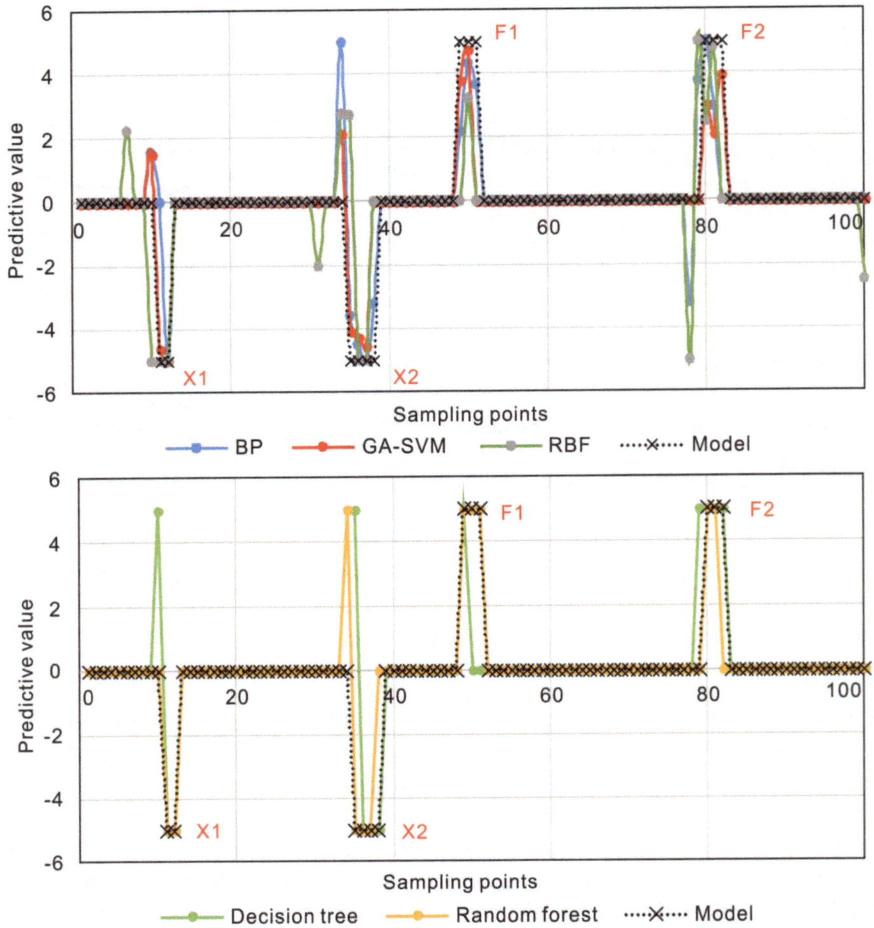


**Fig. 5.8** Prediction results of model data along the coal seam

### *5.1.5 Field Data Prediction*

To further validate the effectiveness of the algorithm, this study utilised 3D coalfield data from Shanxi Province, China, and conducted structural prediction on the dataset. The main coal seams in this area are the No. 8 and No. 15 seams, as shown in Fig. 5.9. The data covers a relatively small area of approximately 1.3 km$^2$ (Fig. 5.9b). Overall, the coal seams dip from north to south, with a reflection time difference of about 50 ms (Fig. 5.9c). The coal-bearing strata belong to the Upper Taiyuan Formation of the Carboniferous System and the Lower Shanxi Formation of the Permian System. The No. 8 coal seam is the target stratum for this study, with a thickness of approximately 5 m and good lateral continuity (Fig. 5.9d). Faults and collapse columns are relatively well developed within the area, with varying scales of development. There are major

**Table 5.3** Comparison of five machine-learning algorithms

| Algorithm | Advantages | Disadvantages | Computation time (s) | Accuracy (%) |
|---|---|---|---|---|
| BP | High classification accuracy<br>Strong fault tolerance<br>Strong memory capability | Requires many parameters<br>Prone to local minima | 3.4 | 94 |
| GA-SVM | Handles high-dimensional problems<br>Strong generalization ability<br>Does not depend on the entire dataset | Low computational efficiency<br>Difficult to implement for large-scale samples | 12 | 96 |
| RBF | High computational efficiency<br>Strong global optimization ability | High requirements on sample quality | 2.9 | 88 |
| Decision tree | Simple computation and easy interpretation<br>No sample preprocessing required | Prone to overfitting<br>Ignores relationships between data | 2.7 | 94 |
| Random forest | Strong resistance to overfitting<br>Strong anti-interference capability | Similar tree structures may occur<br>Small datasets may not yield the optimal classification | 2.8 | 97 |

faults with significant displacement, and three large faults have been identified, with displacements reaching 20–30 m. In order to achieve safe and efficient coal mining, it is essential to carry out precise prediction of the geological structures.

Before coal mining operations commence, roadways must first be excavated, during which certain known structures are exposed. These can be used to validate the results of network-based predictions. Structural interpretation is a complex and time-consuming task; therefore, improving the accuracy and efficiency of interpretation is the objective of this study. This research evaluates the predictive performance of the proposed method through two case studies: 2D planar prediction and 3D sectional prediction.

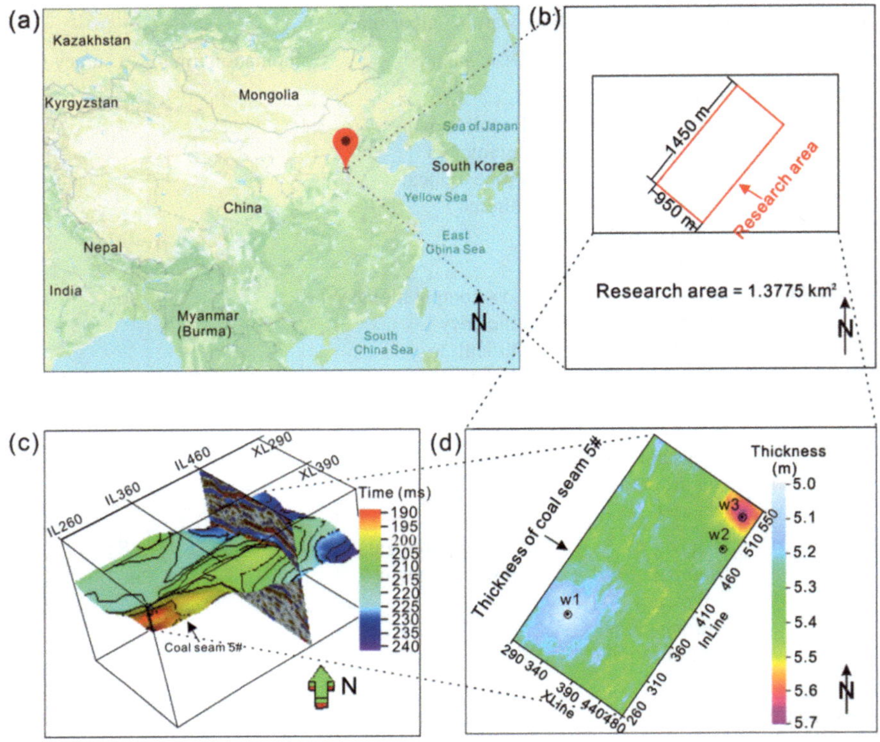


**Fig. 5.9** Field data and overview of the acquisition area

(1) 2D Prediction Along the Coal Seam

A random forest algorithm, known for its high accuracy with simulated data, was employed to achieve 2D planar structural prediction along the coal seam. Firstly, five optimally selected seismic attributes were extracted along the No. 8 coal seam. As shown in Fig. 5.10, planar maps for each of the five attributes and a 3D visualisation of the No. 8 coal seam were generated. Next, to balance the weight of each attribute during model training, all attributes were normalised. Finally, all attribute points (a total of 55,555) on the 2D plane were input into the trained random forest network to obtain the prediction results, as illustrated in Fig. 5.11.

In Fig. 5.11, the black solid line indicates the location of the tunnel. All faults and collapse columns intersecting the tunnel represent exposed structures, with pink denoting predicted collapse columns and purple denoting predicted faults. It can be observed that the prediction results correspond well with both the manually interpreted results and the exposed tunnel findings. At borehole 1, the prediction results are in complete agreement with the tunnel exposure and the manually interpreted extension on both sides; at borehole 2, the predicted fault appears to extend further than indicated by manual interpretation; at borehole 3, there is a certain degree of deviation between the prediction and the manual interpretation; at boreholes 4 and

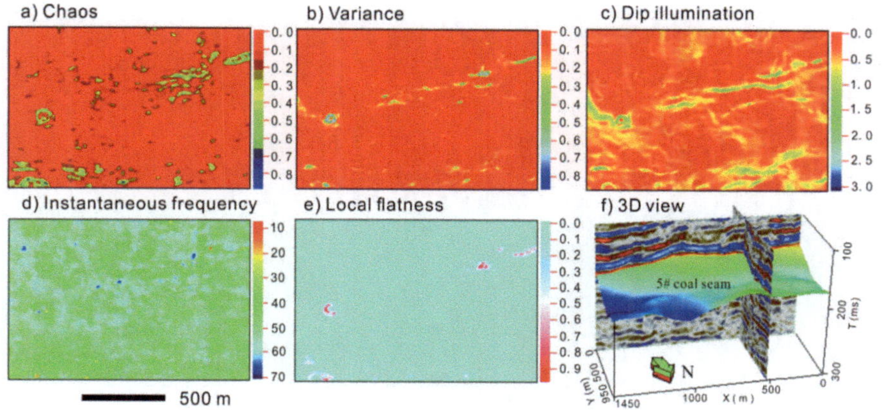


**Fig. 5.10** Seismic attributes extracted along the No. 8 coal seam

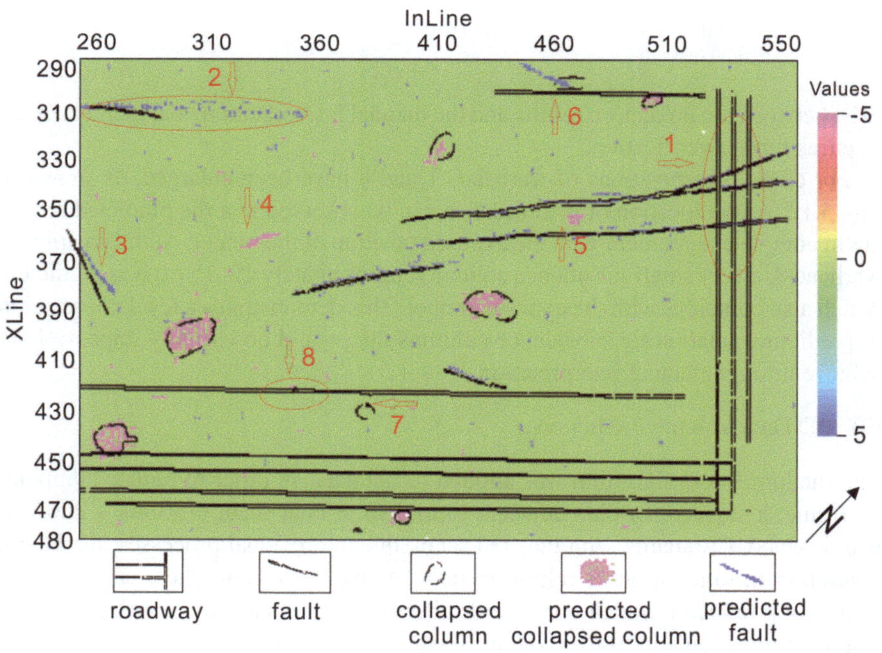


**Fig. 5.11** Geological structure prediction results for the No. 8 coal seam based on the random-forest algorithm

5, two small predicted collapse columns are present, which were not identified in the manual interpretation. From a geological perspective, collapse columns within this study area tend to develop at regular intervals, so the likelihood of these two collapse columns existing is quite high. Boreholes 6 and 7 are locations where discrepancies

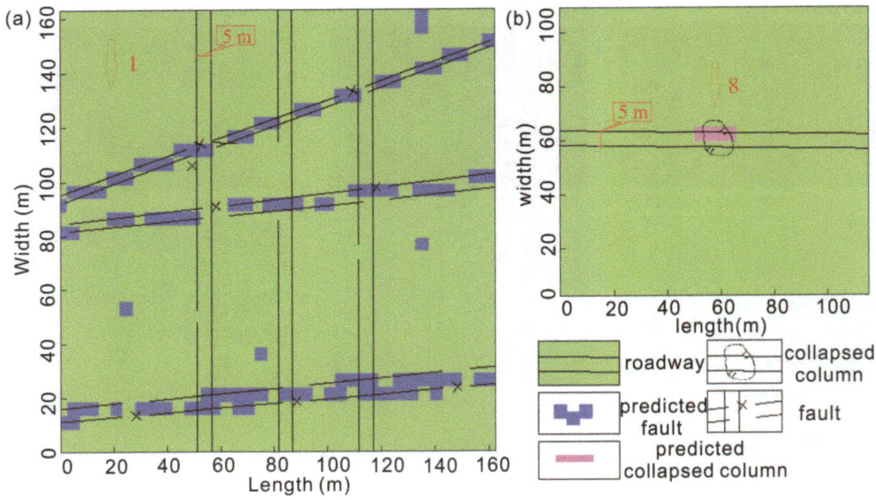


**Fig. 5.12** Local enlargement of the prediction results

exist between the prediction results and the manual interpretation, and their accuracy requires further verification.

For clarity, the positions of warheads 1 and 8 have been enlarged, as shown in Fig. 5.12. At the location of warhead 1, it can be observed that the predicted results are in complete agreement with the exposed position of the tunnel. At the position of warhead 8, a very small subsidence column (approximately 12–15 m) was predicted, which also coincides with the exposed tunnel. This demonstrates a level of precision in predicting small-scale subsidence columns that would be virtually impossible to achieve through manual interpretation.

(2) 3D Data Volume Prediction

The random forest algorithm was applied to 3D data. In order to reduce computation time, a segment of data adjacent to the No. 8 coal seam (150 ms ~ 250 ms) was selected. Compared with the coal seam, due to erosional processes, the lateral continuity of other strata is relatively poor. At the same time, the coal seam acts as a strong reflector, which can shield other strata, and such weak reflections and discontinuities can affect the prediction of structures. Five types of sensitive attribute data volumes were extracted, and predictions were performed on the 3D dataset (a total of 5,611,055 points). The cross-sections of the prediction results are shown in Figs. 5.13 and 5.14, representing the vertical and horizontal survey lines, respectively. It can be seen that the prediction results are in good agreement with the manual interpretation, and some small-scale structures have also been predicted. However, due to the discontinuity of non-coal strata, some predicted structures are not actually caused by real geological structures.

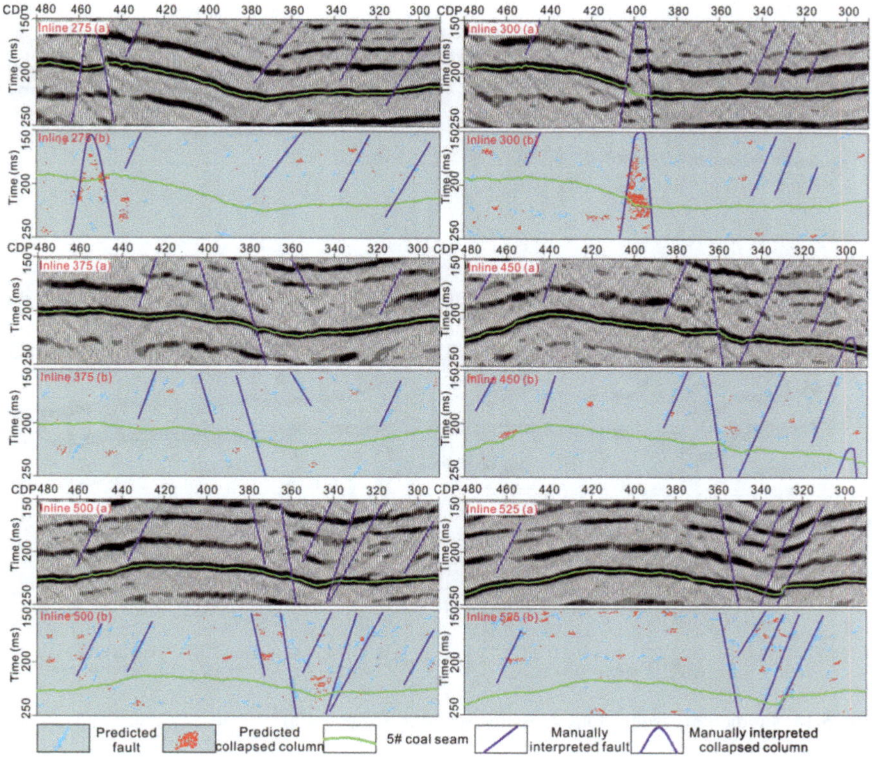


**Fig. 5.13** Prediction results along inline sections 275, 300, 375, 450, 500, and 525

A comprehensive comparison of the two test results for 2D and 3D data demonstrates that both approaches achieve satisfactory predictive outcomes, enabling efficient and high-precision prediction of target geological structures. Naturally, for machine learning-based geological structure interpretation methods, a richer set of training samples can yield more refined predictive results. The 2D planar prediction method offers the following advantages: it is less affected by interference from discontinuities in other strata and exhibits higher computational efficiency. The 3D data prediction method, on the other hand, is beneficial for capturing the morphology, scale, and extension direction of structural development. However, when dealing with large data volumes, its computational efficiency is relatively low, making it less suitable for prediction tasks involving substantial datasets.

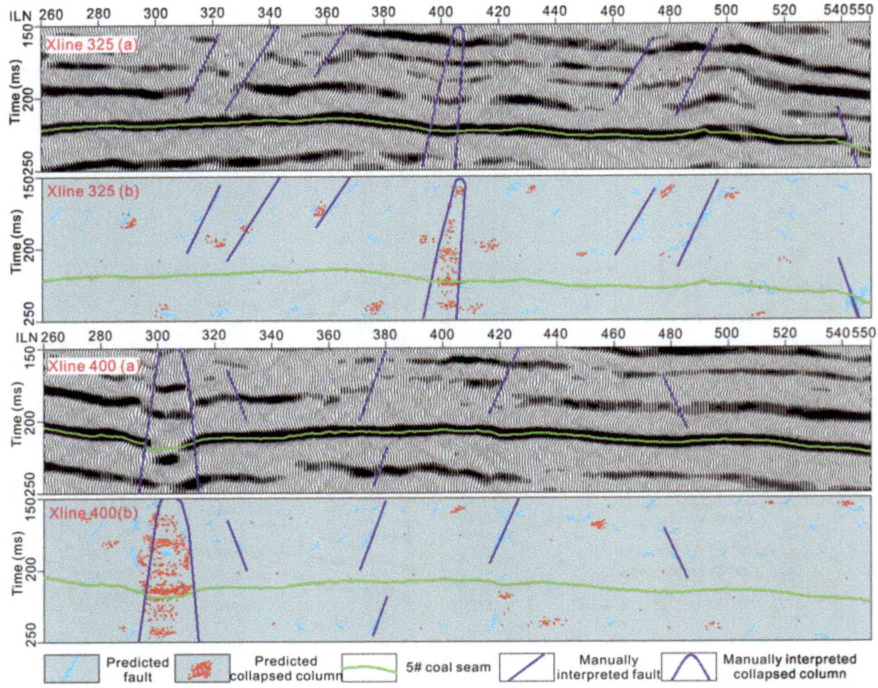


Fig. 5.14 Prediction results along crossline sections 325 and 400

## 5.2 High-Resolution Prediction Technology for Elastic Parameters of Storage Bodies

Full Waveform Inversion (FWI), which makes comprehensive use of both the dynamic and kinematic information contained in the data, is considered one of the most accurate technologies currently available for reservoir parameter prediction. However, the resolution of FWI remains constrained by the seismic frequency bandwidth, and the substantial computational cost severely limits its practical application. To address these issues, this study proposes a seismic full waveform inversion method based on super-resolution (SR-FWI) to enhance both the efficiency and accuracy of reservoir elastic parameter prediction.

### *5.2.1 Constrained Residual Channel Attention Network*

To achieve image super-resolution, the author proposes a Constrained Residual Channel Attention Network (CRCAN) architecture based on regularisation constraints, as illustrated in Fig. 5.15. In addition to the network input of low-resolution images ($\sqrt{m}$) and the output of high-resolution images ($m$), the network primarily comprises two components: feature extraction and data reconstruction. Generally, deeper network structures are more conducive to fine-grained feature extraction. Therefore, in the feature extraction module, shallow feature extraction is performed first, followed by deep feature extraction. This process can be represented as follows:

$$I_f = \mathrm{F}_e(I_{LR}) = F_d(F_s(I_{LR})) \tag{5.40}$$

In the equation, $\mathrm{F}_e$ denotes the feature extraction operation used to obtain the feature map ($I_f$), which is then used as the input data for the data reconstruction module, with $I_f$ serving as the input data. $\mathrm{F}_e$ comprises both shallow feature extraction ($F_s$) and deep feature extraction ($F_d$). For $F_s$, a convolutional layer with $3 \times 3$ is employed, defined by the following equation:

$$I_{sf} = F_s(I_{LR}) = Conv(I_{LR}) \tag{5.41}$$

In the formula, $I_{sf}$ denotes the shallow feature map, and *Conv* represents the convolution operation. For $F_d$, we have added *n* residual channel attention blocks (RCABG) through skip connections, which is beneficial for training very deep convolutional networks. The computation process of the *n*th RCABG can be expressed as follows:

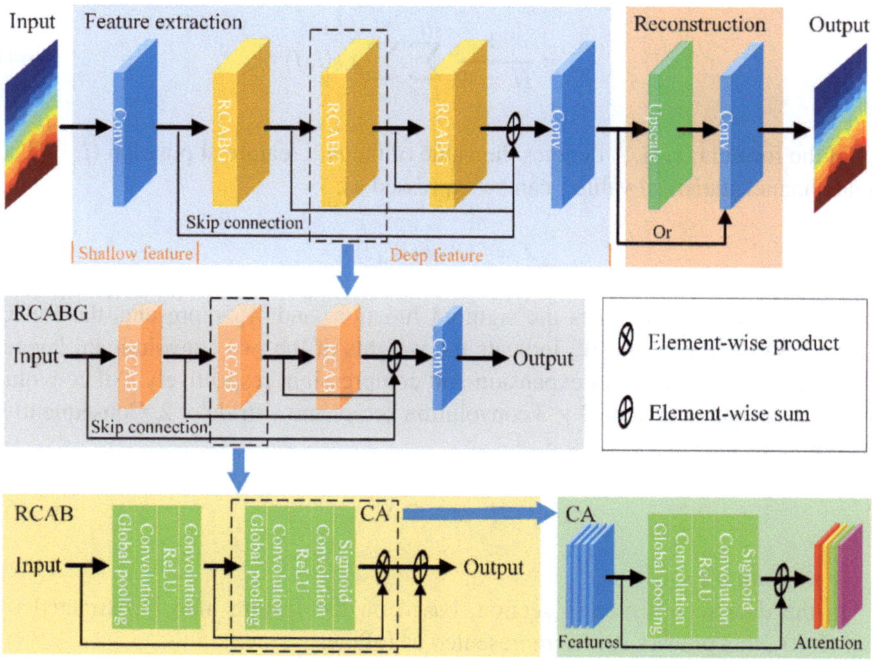


**Fig. 5.15** Architecture of the residual channel-attention network with regularization constraints

$$I_{df\_n} = F_{d\_n}\big(I_{df\_n-1}\big) = F_{d\_n}\big(F_{d\_n-1}(\ldots F_{d\_1}(I_{sf})\ldots)\big) \tag{5.42}$$

In the formula, $F_{d\_n}$ denotes the operation of the $n$th RCABG, while $I_{df\_n-1}$ and $I_{df\_n}$ represent the input and output of the $n$th RCABG, respectively. To prevent the problem of gradient vanishing caused by the simple stacking of RCABGs, skip connections have been introduced. In this case, the extracted deep features $n$ can be expressed as:

$$I_{df} = I_{sf} + W \cdot F_{d\_n}\big(F_{d\_n-1}(\ldots F_{d\_1}(I_{sf})\ldots)\big) \tag{5.43}$$

In the formula, $W$ represents the weight of each RCABG. To better extract low-resolution images, multiple residual channel attention blocks (RCABs) are added within each RCABG. The feature maps extracted from low-resolution images contain information at different scales, each of which has varying importance for image reconstruction. Treating each feature equally is not conducive to high-resolution image reconstruction. Therefore, a channel attention (CA) mechanism is employed to enable the network to focus more on useful information. As shown in Fig. 5.15, each feature channel passes through global average pooling, a fully connected layer (or 1 × 1 convolution), ReLU activation, another fully connected layer (or 1 × 1 convolution), and sigmoid activation to obtain different weights. When the input data 1 × 1 contains 1 × 1 feature maps of size H × W, the statistical value $X$ of the $MH \times W$.

$$z_m = \frac{1}{H \times W}\sum_{i=1}^{H}\sum_{j=1}^{W} x_m(i, j) \tag{5.44}$$

In the formula, $x_m(i, j)$ denotes the value of the $m$th feature at position $(i, j)$. The final channel statistical value $s$ can be expressed as:

$$s = f(W_u \delta(W_d z)) \tag{5.45}$$

In the formula, $f$ denotes the sigmoid function, and $W_d$ represents the ReLU activation function. $\delta$ and $W_u$ indicate the weights of the two convolutional layers, which are used for channel expansion and compression, respectively. All convolutional layers have the same 3 × 3 convolution kernel and a stride of 2. Consequently, the output of the CA, X ~ 3 × 3

$$\tilde{X} = s \odot X \tag{5.46}$$

In the formula, $\odot$.

In the data reconstruction section, based on the feature map $I_f$ extracted by Eq. (5.40), this process can be represented as follows:

$$I_{SR} = F_{RE}\big(F_{UP}(I_f)\big) \tag{5.47}$$

In the equation, $F_{UP}$ denotes the expansion function. The author has employed a transposed convolution, with $F_{RE}$ representing the reconstruction process, utilising a single convolutional layer.

Based on Eqs. (5.40) and (5.47), the computation from low-resolution images to super-resolution images can be achieved. The optimisation of the network is carried out via the loss function, which in this work is optimised by means of structural constraints. The structural information is represented by the tilt angle of the image, which can be calculated through the deviation of the planar wave and is defined as follows:

$$\frac{\partial P(t,x)}{\partial x}+\sigma\frac{\partial P(t,x)}{\partial t}=0 \tag{5.48}$$

In the equation, $P(t, x)$ denotes the plane wave field, $t$ and $x$ represent the two dimensions of the image, and σ indicates the local slope. The conversion between tilt angle and slope is given by the following equation:

$$\theta=\frac{180}{\pi}\tan^{-1}\left(\sigma\frac{\Delta x}{\Delta t}\right) \tag{5.49}$$

In order to incorporate tilt angle information as a constraint term in the loss function, this study constructs a rotation operator, as shown in Fig. 5.16. ($Px$, $Pz$) represents the coordinate system after rotation. The normalised rotation operator can thus be expressed as:

$$P\mathrm{x}=\left[\cos\theta\ \sin\theta\right]\times\begin{bmatrix}FX\\FY\end{bmatrix} \tag{5.50}$$

$$P\mathrm{z}=\left[-\sin\theta\ \cos\theta\right]\times\begin{bmatrix}FX\\FY\end{bmatrix} \tag{5.51}$$

In the equation, $FX$ and $FY$ represent the partial derivatives of the image along the x- and z-directions, respectively.

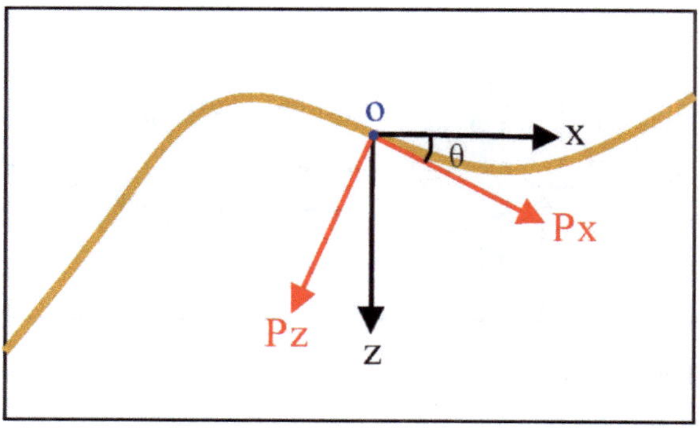

**Fig. 5.16** Construction of the rotation operator

The traditional $L_2$ mean square error loss function can be expressed as:

$$L(\Phi) = \frac{1}{N} \sum_{i=1}^{N} \left\| F_{CRCAN}\left(I_{LR}^{i}\right) - I_{HR}^{i} \right\|_2^2 \tag{5.52}$$

Based on tilt angle information, the improved loss function can be expressed as:

$$L(\Phi) = \frac{1}{N} \sum_{i=1}^{N} \left\| F_{CRCAN}\left(I_{LR}^{i}\right) - I_{HR}^{i} \right\|_2^2 + \lambda_1 \|\mathbb{P}x(\mathbb{I}_{LR})\|_2^2 + \lambda_2 \|\mathbb{P}z(\mathbb{I}_{LR})\|_2^2 \tag{5.53}$$

In the equation, $F_{CRCAN}$ denotes the network parameters, $\Phi$ represents the total number of images, $N$ indicates the nonlinear mapping relationship from IR to HR images established by the CRCAN proposed in this study, and λ1 and λ2 denote the regularisation parameters.

### 5.2.2 SR-FWI Implementation Process

The objective of this study is to enhance the computational efficiency and accuracy of full waveform inversion (FWI) based on super-resolution (SR) networks. As illustrated in Fig. 5.17, starting from the initial model, one or several iterations of FWI are first performed to output an intermediate result, rather than outputting the result after completing all iterations. Next, this intermediate result is used as a low-resolution image and input into the CRCAN network to obtain a super-resolution image. The obtained super-resolution image is then used as the initial model for a subsequent round of FWI. This process is repeated iteratively until the desired requirements are met, at which point the final high-resolution result is output. During this iterative process, the neural network serves to regularise the FWI, meaning that SR-FWI can be regarded as an implicit regularisation method, which helps to alleviate the ill-posedness of the inversion problem.

The author employed time-domain elastic full waveform inversion (FWI), with the relevant computational procedure described as follows. FWI iteratively computes the

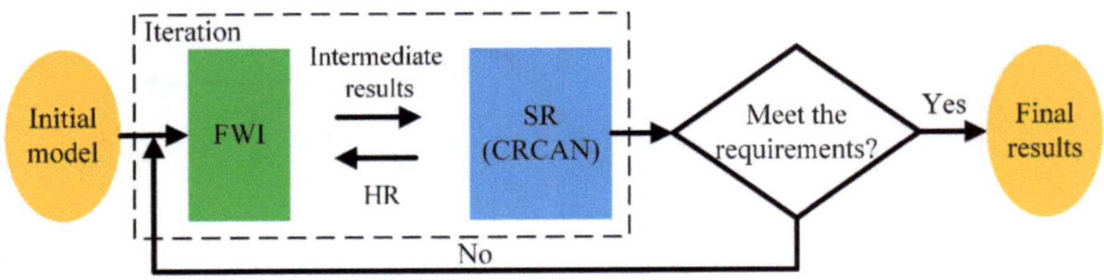


**Fig. 5.17** Workflow of full-waveform inversion based on a super-resolution network (SR-FWI)

optimal solution for subsurface parameters, typically by minimising a least-squares objective function, which can be expressed as:

$$E(m) = \frac{1}{2}\sum_{\mathrm{ns}}\sum_{nr}\int_{0}^{T}\left\|d^{obs} - G \cdot m\right\|^{2} dt \tag{5.54}$$

In the equation, $d^{obs}$ represents the observed seismic data, $G$ denotes the synthetic seismogram, $m$ stands for the model parameters, $ns$ is the number of shot points, $nr$ is the number of receiver points, and $T$ indicates the recording time. Forward modelling is a key factor affecting the accuracy of inversion. In this study, a spatial fourth-order and temporal second-order staggered grid scheme is employed to simulate the seismic records. The wave equation under the assumption of isotropic homogeneous media in all directions can be written as:

$$\rho\frac{\partial v_i}{\partial t} = \delta_{ij,j} + f_i \tag{5.55}$$

$$\frac{\partial \delta_{ij}}{\partial t} = \frac{1}{2}E_{ijkl}\left\{v_{k,l} + v_{l,k}\right\} \tag{5.56}$$

In the equation, $f_i$ denotes the density, $\rho$ represents the velocity component, δij indicates the stress tensor, $v_i$ denotes the external force vector, and $\delta_{ij}$ represents the elastic modulus.

The objective of FWI is to update the model parameters $m$ by minimising the objective function (Eq. 5.54), which can be expressed as follows:

$$m = m_0 + \Delta m = m_0 + \alpha \cdot Gr \tag{5.57}$$

In the equation, $m_0$ represents the initial model, $Gr$ denotes the update at each iteration, $Gr$ is the step size, and $\Delta m$ is the gradient of the objective function. $\alpha$ can be calculated using the following formula:

$$\frac{\partial E(m)}{\partial m} = \sum_{ns}\sum_{nr}\int_{0}^{T}\left[\left(\frac{\partial (G \cdot m)}{\partial m}\right) \cdot \left(d^{obs} - G \cdot m\right)\right] dt \tag{5.58}$$

This gradient can be obtained by calculating the forward and backward propagated wave fields.

Based on the above algorithm, when applying SR-FWI to field data, Algorithm 1 can be used for computation.

**Algorithm 1** Full-waveform inversion method based on CRCAN

CRCAN network training

Input: low-resolution images, high-resolution images, number of iterations, size of low-resolution images, size of high-resolution images, batch size, learning rate, number of training samples $\{I_{LR}\}_{n=1}^{N}$ $\{I_{HR}\}_{n=1}^{N}$ $Iter$ $size_{LR}$ $size_{HR}$ $bs$ $lr$ $N$

Initialization: $loss = 0$ $I_{SR}^{0} = \{I_{LR}\}$

for to do $i = 1$ $Iter$

for to do $n = 1$ $N/bs$

$I_{SR} = F_{CRCAN}(I_{LR}) = F_{RE}(F_{UP}(F_d(F_s(I_{LR}))))$; // computation for the input data $F_{(\cdot)}$

$loss = L_2(I_{SR}, I_{HR})$;

end

$\Theta_{n+1} = Adam(\Theta_n; loss; lr)$; // network parameter set $\Theta$

End

SR-FWI procedure

Input: observed shot-gather seismic data, source and receiver locations, initial model including P-wave velocity, S-wave velocity, and density, established network for image super-resolution, number of iterations, number of cycles $d^{obs}$ $SPS$ $m_0$ $CRCAN$ $It$ $Nc$

for to do $i = 1$ $Nc$

for to do // 10 is a constant and can be modified according to the data $j = 1$ $It/10$

$d^{cal} = FD(m_0)$; // denotes elastic-wave forward modelling $FD$

$loss = L_2(d^{obs}, d^{cal})$;

$m = m_0 + \Delta m$;

end

$I_{SR} = CRCAN(I_{LR}; \Theta)$; // denotes the established nonlinear mapping $CRCAN$

$m_0 = I_{SR}$;

end

Output: high-resolution reservoir P-wave velocity, S-wave velocity, and density

### 5.2.3 *Preparation of Training Data and Network Training*

Image super-resolution is applicable across a wide range of fields. For seismic inversion tasks, the author has generated random geological models and combined these geological models with the DIV2K dataset for network training. The construction of 2D geological models primarily involves the following four steps:

(1) Establishment of a horizontal layered geological model. Given the model dimensions, within a certain range, the number of geological layers, the thickness of each layer, and the velocity of each geological layer are randomly assigned. The resulting horizontal geological model is shown in Fig. 5.18a.
(2) Simulation of folded geological strata. Based on the layered geological model, the structure of folded strata was simulated, and the results obtained are shown in Fig. 5.18b.

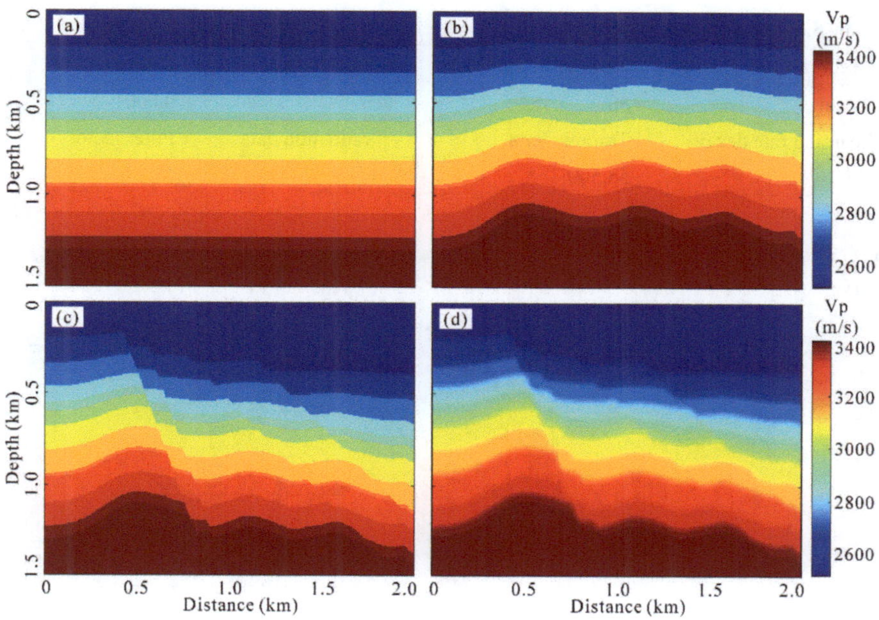


**Fig. 5.18** Construction of randomly generated geological models

(3) Simulation of faults. Based on the geological model with fold structures, planar shearing is introduced to simulate the development of faults. The resulting outcome is shown in Fig. 5.18c.

(4) By following steps (1)–(3), a large number of high-resolution random geological model images (HR) can be obtained. To facilitate the training of the super-resolution network, low-resolution images (LR) can be generated using an averaging filter method. By applying multiple filter kernels, LR images with varying degrees of blurring can be produced. The result for a filter kernel size of 15 is shown in Fig. 5.18d.

Based on the aforementioned method, 500 high-resolution random geological models (with dimensions of 2000 × 2000) were obtained. Combined with 800 high-resolution natural images from the DIV2K dataset, these together formed the HR dataset. Using mean filtering methods with different filter kernels, a corresponding LR dataset of 1300 images was generated. Subsequently, for each image, 50 patches of size 200 × 200 were randomly cropped, resulting in a total of 65,000 data pairs containing both LR and HR images. Of these, 90% were used for network training and 10% for network validation. The parameters used in network training are shown in Table 5.4, where the filter kernel denotes the parameter used to obtain training data with different degrees of blurring at low resolution; this process is completed during the generation of the training dataset. The learning curve of the network is shown in Fig. 5.19, from which it can be seen that as the number of iterations increases, the loss function value (Loss) gradually decreases, while the prediction accuracy (peak

**Table 5.4** Network training parameters

| Optimizer | Number of iterations | Batch size | Learning rate |
|---|---|---|---|
| Adam | 50 | 16 | 0.0001 |
| Learning rate decay | Training data | Validation data | Filter factor |
| 0.5 | 58,500 | 6500 | 10/15/25 |

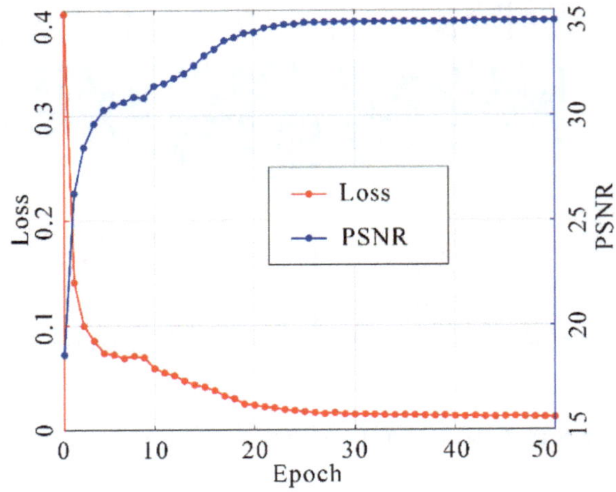


**Fig. 5.19** Network learning curves

signal-to-noise ratio, PSNR) gradually increases. After 20 iterations, the results tend to stabilise, indicating that the network has converged.

### 5.2.4 *Network Prediction*

The author evaluated the predictive performance of the network using two geological models and one natural image that were not involved in network training, as shown in Fig. 5.20. The first column displays the original high-resolution images (HR), the second column shows the low-resolution images (LR) obtained via bicubic filtering, and the third column presents the super-resolution images (SR) predicted by the network. The first row represents a typical geological model containing faults; the second row features an image with an "ostrich" pattern, which can be used to represent complex reservoir parameter variations, and Gaussian noise with SNR = 40 dB was added to the image. The third row depicts a geological model with local anomalies, which can represent time-varying reservoir conditions, and Gaussian noise with SNR = 25 dB was added to the image. The image noise is used to test the noise robustness of the network. To demonstrate the generalisation capability of the network, the structural similarity index (SSIM) between the three selected models and all training models was calculated. The SSIM values for (a), (d), and (g) are 0.28, 0.17, and 0.25, respectively, indicating sufficient differences between the test data and the training data. From the prediction results, it can be seen that the proposed CRCAN network can effectively enhance image resolution. At the same time, the SSIM of the predicted results is shown in the upper right corner of each image. It can be observed that the quality of the predicted images is greatly improved. Therefore, this method can finely delineate geological boundaries and small-scale local anomalies.

In order to compare the computational performance of the objective function based on structural constraints with that of the traditional objective function (without constraint terms), the author takes Fig. 5.21a as an example to test the prediction results under both scenarios. Figure 5.21b shows the low-resolution image with a filter kernel of 20, while Fig. 5.21c, d present the prediction results with and without structural constraints, respectively. From the position of the arrowhead in the figures, it can be observed that the improved objective function is beneficial for the high-precision delineation of subsurface boundaries, a conclusion further supported by the SSIM values in the upper right corner. In addition, the loss curves of the

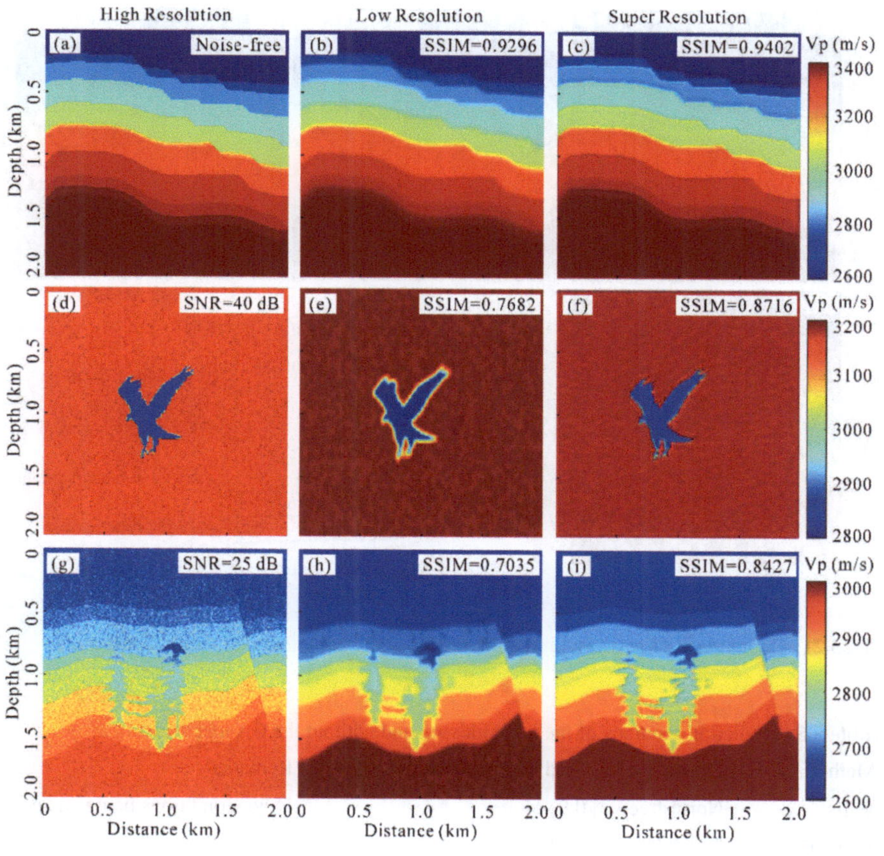


**Fig. 5.20** Prediction results of the CRCAN network

objective function residuals (Loss) with respect to the number of iterations were also summarised, as shown in Fig. 5.21e, where the blue and red lines represent the residual curves of the traditional and improved objective functions, respectively. It can be seen that, under structural constraints, the residuals of the objective function after stabilisation are reduced, indicating that the structural constraints enhance the fit between the predicted results and the original data, thereby improving the prediction accuracy of the network.

In addition, to illustrate the contribution of the CA mechanism to the network, the author also compares the prediction accuracy of the network with and without CA, taking Fig. 5.21a as an example. This comparison is conducted without any structural constraints. The prediction accuracy results are shown in Table 5.5. The results indicate that CA can effectively improve the prediction accuracy of the network. Although the SSIM value increases by only about 0.01, it should be noted that further improving accuracy on the basis of already high-precision prediction results is challenging, and even a small improvement is valuable.

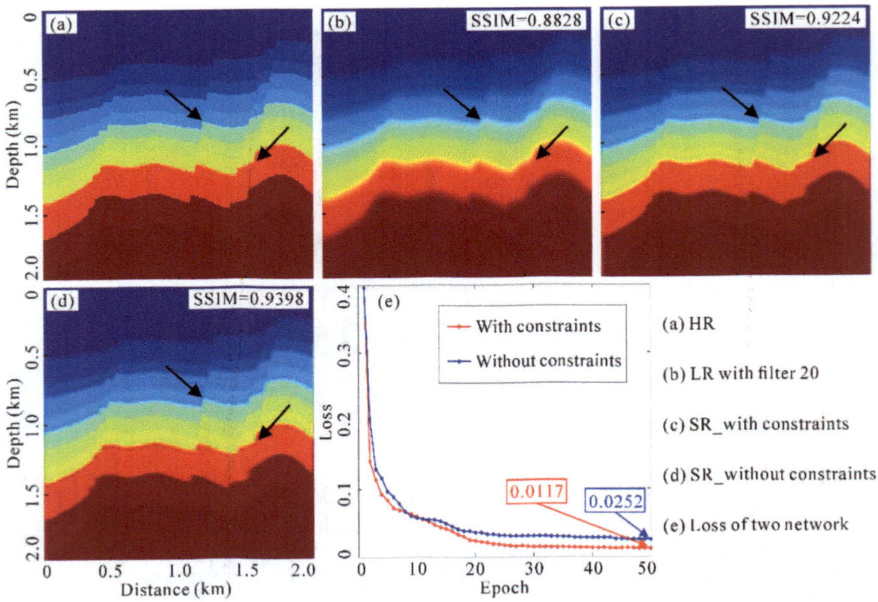


**Fig. 5.21** Comparison of network prediction results with and without structural constraints

**Table 5.5** Prediction accuracy of the network with and without the CA mechanism

| Method | Data noise | Prediction accuracy (SSIM) | Remarks |
|---|---|---|---|
| With CA | Noise-free | 0.9224 | Other conditions being equal |
| Without CA | Noise-free | 0.9127 | |

### 5.2.5 Synthesis Data Testing and Analysis

The author employed synthetic data to test the effectiveness of CRCAN when applied to full waveform inversion (SR-FWI). The time-domain elastic wave full waveform inversion method was used to validate the proposed workflow, with the LBFGS optimisation algorithm adopted for model updating. Both the forward and inverse modelling processes were conducted within a 400 (horizontal) × 400 (vertical) grid with a spacing of 5 m. The seismic source was a Ricker wavelet with a dominant frequency of 20 Hz, and the recorded data comprised the two components of particle velocity. Taking $CO_2$ reservoir monitoring in geological storage as an example, a geological model with local anomalies was constructed to simulate various plume morphologies of $CO_2$, and the initial model was obtained by smoothing the true model. Figure 5.22 shows the true model and the corresponding initial models for P-wave velocity, S-wave velocity, and density. The inversion results of the first two models based on SR-FWI are presented in Fig. 5.23, displaying, from left to right, the results for P-wave velocity, S-wave velocity, and density. It can be observed that SR-FWI effectively recovers the layer boundaries and geological anomalies. The calculated SSIM indicates that the inversion results achieve high accuracy.

To compare the computational performance of SR-FWI, the author conducted calculations on synthetic data using both traditional FWI (TFWI) and total variation regularised FWI (TV-FWI). As shown in Fig. 5.24, which presents the comparative results for the model in Fig. 5.23c, panels (a), (b), and (c) correspond to the inversion results of TFWI, TV-FWI, and SR-FWI, respectively. From the arrowhead region in the figures, it can be observed that, compared with TFWI and TV-FWI, SR-FWI provides a clearer prediction of small-scale features such as boundaries, thereby improving resolution to a certain extent. The calculated SSIM values also support

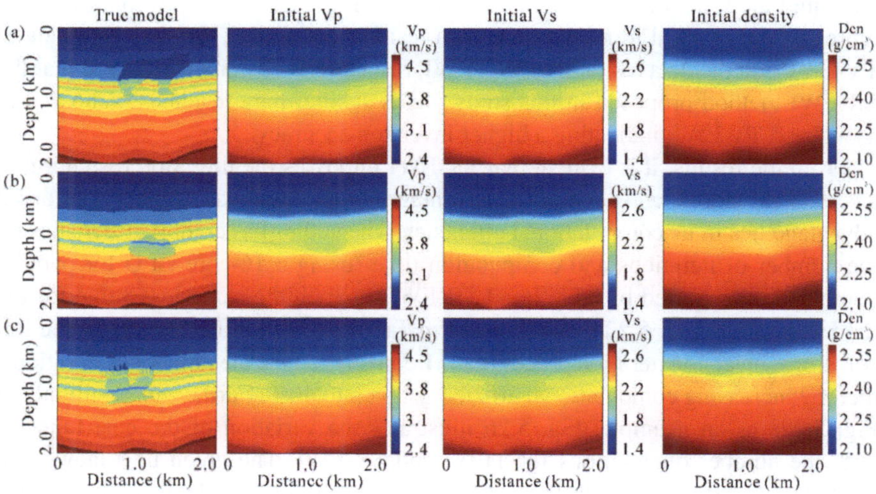


**Fig. 5.22** Geological models used for testing and their corresponding initial models

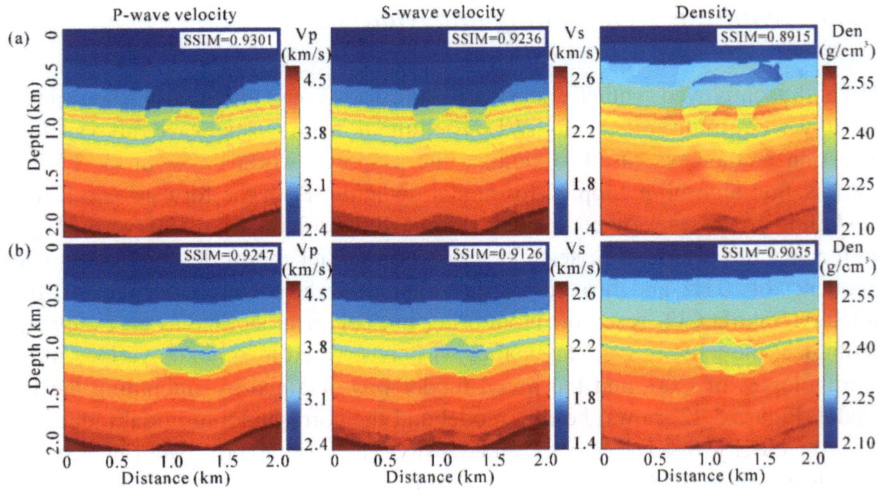


**Fig. 5.23** Inversion results based on SR-FWI

this conclusion. To clearly illustrate the differences among the three inversion results, the inversion profiles at the virtual line position were further extracted, as shown in Fig. 5.24d. Here, the black line represents the true model, the green line the TFWI result, the blue line the TV-FWI result, and the red line the SR-FWI result. In addition, the Pearson correlation coefficients were calculated for the region indicated by the dashed box in the figure. The comparison shows that, relative to TFWI and TV-FWI, SR-FWI achieves higher resolution of geological layer or anomaly boundaries, with the effect being particularly pronounced at the arrowhead region.

In addition, a comparison was made of the data residuals for the three methods, including the x- and z-components, as shown in Fig. 5.25. From left to right, these represent the original input data, the data residuals of TFWI, the data residuals of TV-FWI, and the data residuals of SR-FWI, respectively. It can be observed that the SR-FWI method proposed in this study yields smaller data residuals compared to TFWI and TV-FWI, indicating a higher inversion accuracy.

From the results, it is evident that SR-FWI improves the inversion resolution to a certain extent, which is one of the advantages of this method. Another significant advantage lies in its computational efficiency. The author has compiled statistics on the number of iterations and computation time (using a 16-core CPU for parallel computation) required by TFWI and SR-FWI to reach the same loss value for the model shown in Fig. 5.22a, as presented in Table 5.6. Applying SR during the full waveform inversion iteration process can effectively reduce the number of iterations, thereby increasing computational efficiency by up to 40%. Furthermore, still taking Fig. 5.22a as an example, Fig. 5.26 illustrates the variation of the cost function with the number of iterations for TFWI and SR-FWI. The green line shows that the CRCAN network, which is used for super-resolution, accelerates convergence during inversion, enabling SR-FWI to achieve a faster convergence rate compared

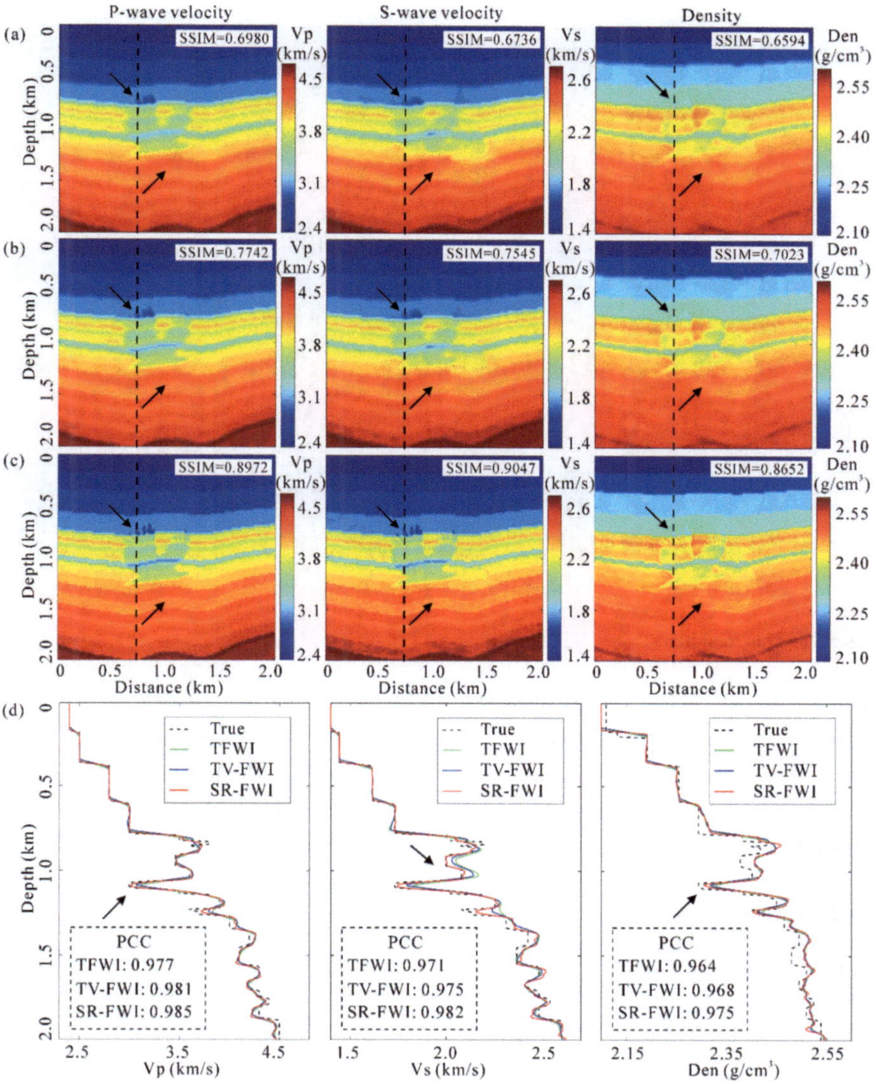


**Fig. 5.24** Inversion results based on TFWI, TV-FWI, and the SR-FWI proposed in this study

to TFWI. In this study, SR is introduced once every three iterations, resulting in six SR operations corresponding to the green line in the figure, five of which effectively reduce the residual between the simulated and observed data. In the remaining SR operation, although the residual increases, it remains under control.

In addition, the inversion accuracy under different data noise levels was also tested, as shown in Table 5.7. Regardless of whether traditional FWI or SR-FWI was used, the inversion accuracy improved with increasing signal-to-noise ratio. However, SR-FWI demonstrated better predictive performance.

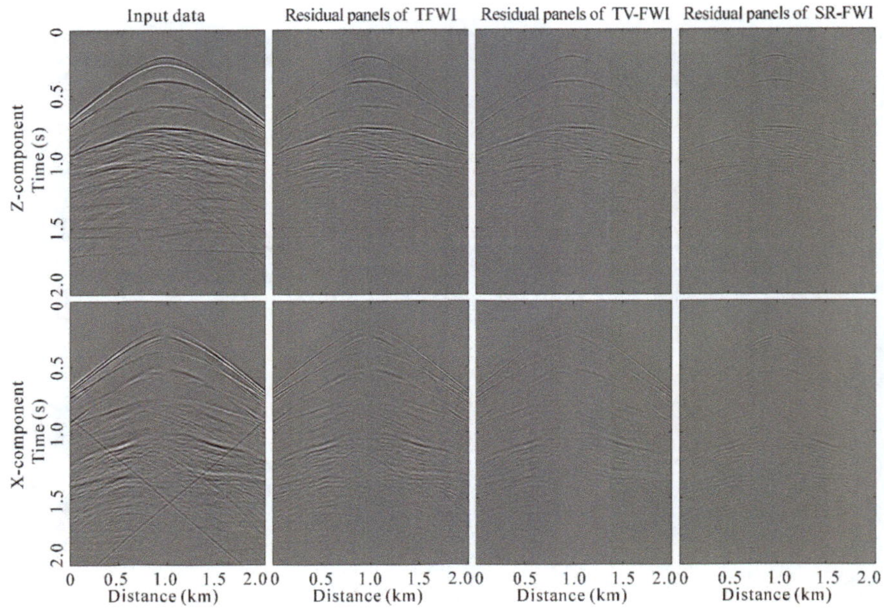


**Fig. 5.25** Data residuals

**Table 5.6** Computation time of TFWI and SR-FWI at the same loss value (0.05)

| Method | Number of iterations | | Training time (h) | Computation time (h) | |
|---|---|---|---|---|---|
| | 20 shots | 100 shots | | 20 shots | 100 shots |
| TFWI | 86 | 82 | / | 13.27 | 41.68 |
| SR-FWI | 62 | 38 | 27.4 | 8.9 | 23.81 |

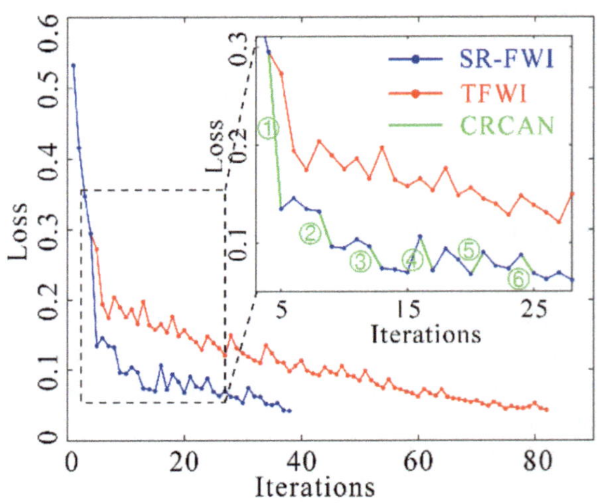


**Fig. 5.26** Loss-function curves for TFWI and SR-FWI

**Table 5.7** Inversion accuracy at different noise levels

| Noise level | TFWI (SSIM) | SR-FWI (SSIM) |
|---|---|---|
| SNR = 20 | 0.62 | 0.79 |
| SNR = 25 | 0.67 | 0.85 |
| SNR = 30 | 0.71 | 0.88 |
| No noise | 0.74 | 0.92 |

### 5.2.6 Field Data Application

The author conducted field tests of the proposed SR-FWI method using data from a walkaway VSP seismic line in a mining area in western China. As the target coal seam is relatively shallow, the data recording time was 2 s, with a sampling interval of 0.5 ms. As shown in Fig. 4.29a, the shot line is 1000 m long with a spacing of 10 m, and the DSU3 receivers are positioned in the borehole from 20 to 970 m at 10 m intervals, recording velocity information. Due to data quality considerations, only the vertical component (z-) of the field data was used. To match the 2D FWI proposed in this study, a 3D-2D conversion was performed on the data. This can, to some extent, mitigate frequency and phase errors caused by the 2D approximation. The inversion target area is indicated by the square frame in the figure. The initial inversion model was a low-frequency model constructed using well log curves constrained by stratigraphic horizons and then smoothed, with the initial models of P-wave velocity, S-wave velocity, and density shown in Fig. 5.27a. Under the same conditions, the inversion result of traditional FWI (TFWI) is shown in Fig. 5.27b, while the inversion result of the proposed SR-FWI is shown in Fig. 5.27c. It can be observed that, based on the high-resolution VSP seismic data, TFWI also achieves relatively good inversion results, especially for P-wave velocity. However, as indicated by the arrows in the figure, the inversion result of the proposed SR-FWI achieves higher resolution in some thin reservoir layers compared to TFWI.

For clarity of comparison, Fig. 5.28 presents the curves of two inversion results at the well location. The black solid and dashed lines represent the actual well log curves and the initial model, respectively, while the green and red lines correspond to the inversion results of TFWI and SR-FWI, respectively. From left to right, the figure displays the P-wave velocity, S-wave velocity, and density. Additionally, the Poisson's ratio correlation coefficient has been calculated and is shown within the dashed box in the figure. The arrows in the figure indicate that SR-FWI achieves superior resolution and prediction accuracy in thin reservoir layers. The above analysis of the curves demonstrates that the SR-FWI method proposed by the author aligns more closely with the well log curves compared to TFWI. Therefore, the inversion results of SR-FWI exhibit higher accuracy and resolution.

In addition, the data residuals were also calculated, as shown in Fig. 5.29, which presents the data residual sections for two different offset distances (Fig. 5.29a with an offset of 30 m, and Fig. 5.29b with an offset of 450 m). From left to right, the panels display the input vertical component shot gather data, the residual section from

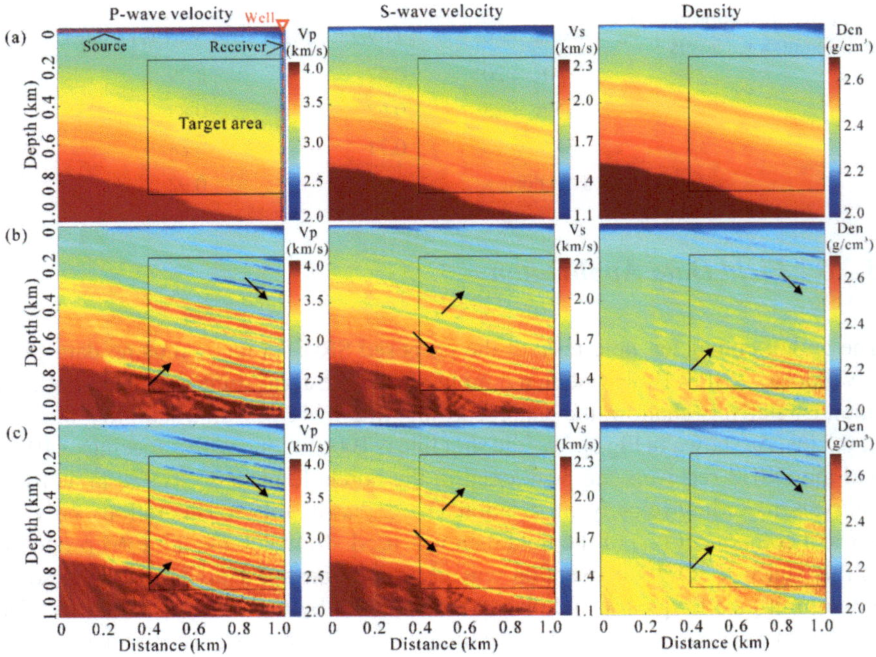


Fig. 5.27 Initial inversion model and inversion results for the field data

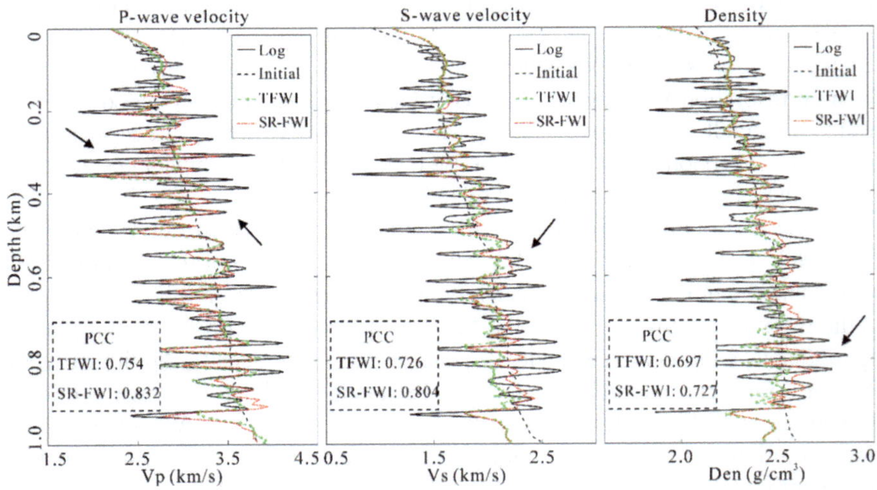


Fig. 5.28 Curves of inversion results at the well location

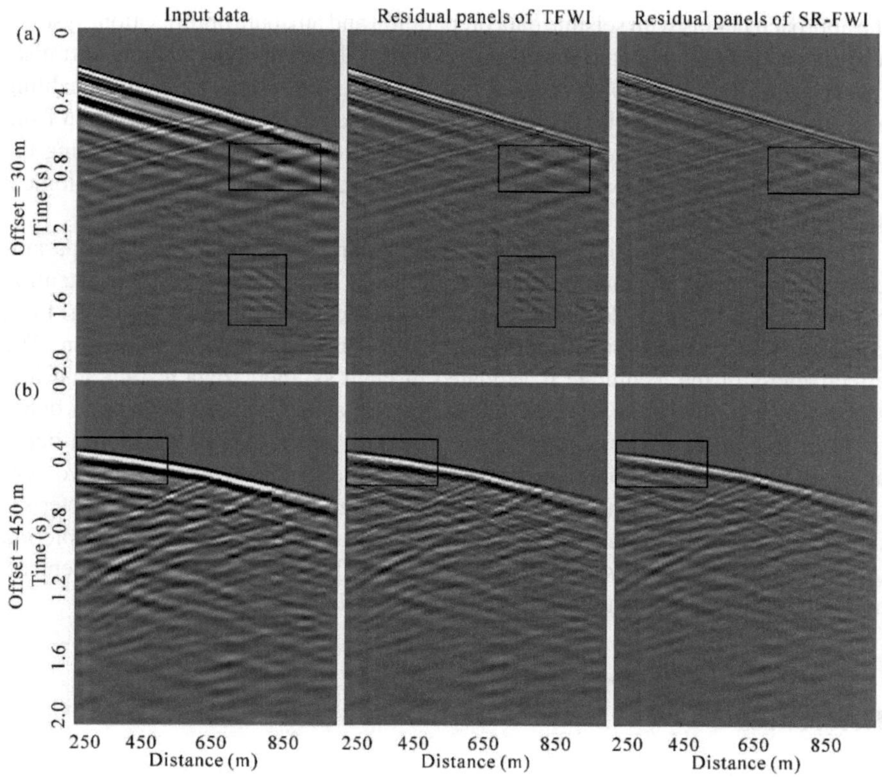


**Fig. 5.29** Data residual panels

TFWI, and the residual section from SR-FWI, respectively. The results indicate that, although the residuals are relatively large compared to those from synthetic data, both methods achieve a good match for the main waveform positions. Furthermore, SR-FWI further reduces the data residuals at certain locations (highlighted by the boxes in the figure).

## 5.3 Section Summary

This chapter focuses on advanced characterisation techniques for $CO_2$ geological storage reservoirs, establishing an intelligent, high-resolution reservoir characterisation system. This system provides core technical support for the scientific site selection, effective monitoring, and precise safety assessment of $CO_2$ geological storage. Addressing the need for intelligent identification of reservoir geological structures, the approach begins with seismic data preparation and attribute optimisation, systematically introducing machine learning algorithms. Through model training optimisation and algorithm selection processes, performance is enhanced, ultimately enabling accurate prediction of real-world data. This forms a complete technical chain from data pre-processing to intelligent structural identification, effectively resolving the industry challenges of low efficiency and limited accuracy in traditional geological structure identification.

In response to the demand for high-resolution characterisation of the elastic parameters of storage formations, this chapter proposes a regularised residual attention network model. This model, combined with a dedicated implementation workflow, completes the preparation of training data, network training, and prediction. The effectiveness of the technique is validated through synthetic data testing, and it is successfully applied to real-world data scenarios, achieving high-resolution quantitative description of elastic parameters. This provides a reliable tool for in-depth analysis and dynamic monitoring of reservoir properties. By constructing two core technical systems, this chapter overcomes the limitations of traditional characterisation methods, establishing a robust technical foundation for the fine-scale depiction of $CO_2$ storage formations, and thereby enhancing the scientific rigour, refinement, and long-term safety assurance of storage engineering.

## References


1. Li D, Peng SP, Lu YX et al (2019) Seismic structure interpretation based on machine learning: a case study in coal mining. Interpretation J Subsurf Charact 7:SE69–SE79
2. Li D, Guo Y, Peng SP et al (2023) Seismic full-waveform inversion based on super-resolution for high-precision prediction of reservoir parameters. Geophysics 88(3):R243–R256
3. Li H, Jian H, Wei W (2018) The sustainable development assessment of reservoir resettlement based on a BP neural network. Int J Environ Res Public Health 15(1):146–161
4. Wang X (2012) Quantitative evaluation of the development patterns and complexity of the Shenhe No. 2 mine structure. Xi'an University of Science and Technology, Xi'an
5. Zhang X (2018) Research on extraction and classification methods of typical geological structure features in elastic wave echo signals. Shandong University of Science and Technology, Qingdao
6. Zhang Y (2020) Research on profit forecasting of the big health industry based on RBF neural network. Nanjing University of Posts and Telecommunications, Nanjing
7. Rostami A, Arabloo M, Lee M et al (2018) Applying SVM framework for modeling of $CO_2$ solubility in oil during $CO_2$ flooding. Fuel 214:73–87
8. Routray S, Ray AK, Mishra C et al (2018) Efficient hybrid image denoising scheme based on SVM classification. Optik 157:503–511
9. Zhou Y (2018) Research on early warning of spontaneous combustion of residual coal in goaf of high-gas prone self-ignition coal seams. Anhui University of Science and Technology, Anhui
10. Jiao LB (2011) Application of genetic algorithm-support vector machine (GA-SVM) for damage identification of bridge. Int J Comput Intell Appl 10(4):383–397
11. Marjanović M, Krautblatter M, Abolmasov B et al (2018) The rainfall-induced landsliding in Western Serbia: a temporal prediction approach using decision tree technique. Eng Geol 232:147–159
12. Wang K (2020) Research on P2P pre-loan credit risk assessment method based on improved random forest algorithm. Nanjing University of Posts and Telecommunications, Nanjing

13. Breiman L (2001) Using iterated bagging to debias regressions. Mach Learn 45:261–277
14. Dong L, Li X, Xie G (2014) Nonlinear methodologies for identifying seismic event and nuclear explosion using random forest, support vector machine, and naive Bayes classification. Abstr Appl Anal 12:1–8
15. Hibert C, Michéa D, Provost F et al (2017) Automated seismic detection of landslides at regional scales: a random forest based detection algorithm. In: AGU fall meeting abstracts
16. Fomel S (2002) Application of plane-wave destruction filters. Geophysics 67:1946–1960
17. Li D, Peng SP, Zhang R et al (2021) Prestack seismic inversion with structural constraints. Interpretation J Subsurf Charact 9:T495–T506
18. He Q, Wang Y (2021) Reparameterized full-waveform inversion using deep neural networks. Geophysics 86(1):V1–V13
19. Zhu W, Xu K, Darve E et al (2022) Integrating deep neural networks with full-waveform inversion: reparameterization, regularization, and uncertainty quantification. Geophysics 87:R93–R109
20. Köhn D, De ND, Kurzmann A et al (2012) On the influence of model parametrization in elastic full waveform tomography. Geophys J Int 191(1):325–345
21. Agustsson E, Timofte R (2017) NTIRE 2017 challenge on single image super-resolution: dataset and study. In: IEEE conference on computer vision and pattern recognition workshops, vol 7, pp 126–135
22. Li D, Peng S, Guo Y et al (2021) $CO_2$ storage monitoring based on time-lapse seismic data via deep learning. Int J Greenh Gas Control 108(1):103336

# Chapter 6
# Geological Storage and Migration Monitoring Technologies for Carbon Dioxide

$CO_2$ storage is a dynamic process, involving the continuous injection of $CO_2$ into geological formations over a certain period. However, the migration and dispersion behaviour of $CO_2$ underground remains uncertain. At present, time-lapse seismic monitoring is the primary method used to monitor the extent of $CO_2$ migration.

While oil and gas are being displaced, $CO_2$ will be sequestered underground. However, two key issues arise: firstly, how to ensure that $CO_2$ is securely stored without leakage, including diffusion into adjacent saline aquifers, surface aquifers, or the atmosphere; and secondly, how to promptly detect regions where $CO_2$ has not yet reached, so as to adjust oil or gas displacement strategies in a timely manner. The continuous injection of $CO_2$ leads to fluid substitution within the reservoir, resulting in changes to the reservoir's elastic parameters. This provides a theoretical basis for monitoring the migration range of $CO_2$ using seismic methods. Time-lapse seismic monitoring is the most effective technology for tracking the migration and storage volume of $CO_2$. By comparing seismic data acquired before and after $CO_2$ injection, as well as at different stages of injection, it is possible to determine the subsurface distribution of $CO_2$ at various stages. However, the changes in reservoir parameters induced by $CO_2$ injection are subtle, thus requiring high-precision predictive methods to characterise the migration of $CO_2$.

Deep learning, as a type of deep neural network architecture, can extract more information from data under data-driven conditions and can directly establish the mapping relationship from seismic data to reservoir elastic parameters with high efficiency. Therefore, it is feasible to consider using deep learning to predict the migration behaviour of $CO_2$. In this study, a deep learning-based method is proposed to establish a nonlinear mapping relationship between time-lapse seismic data and variations in reservoir parameters. Moreover, variations in seismic reflection wave amplitudes typically represent the combined response of multiple reservoir parameters, and it is often difficult to distinguish the contribution of each parameter. To provide a more detailed description of $CO_2$ migration states, this study investigates a method for simultaneous prediction of multiple reservoir parameters, simulating



D. Li and Y. Guo, *Geological Storage and Monitoring of Carbon Dioxide*,
https://doi.org/10.1007/978-981-92-3621-3_6

the changes in reservoir parameters during the $CO_2$ injection process, and thereby achieving $CO_2$ migration monitoring based on reservoir parameters.

## 6.1 Carbon Dioxide Plume Rapid Imaging Technology

Deep learning algorithms can directly establish a nonlinear mapping between seismic data and reservoir elastic parameters under data-driven conditions. Once the deep learning network has been properly trained, it can rapidly predict reservoir elastic parameters using seismic data, typically completing the process within a few seconds. Therefore, this study proposes a $CO_2$ plume rapid imaging method based on a fully convolutional neural network (FCN), which constructs the relationship between time-lapse seismic data and variations in time-lapse reservoir parameters, and applies this approach to $CO_2$ storage monitoring.

### *6.1.1 Fully Convolutional Networks*

Fully Convolutional Networks (FCNs) are derived from Convolutional Neural Networks (CNNs) by replacing the fully connected layers with convolutional layers, thereby removing the restriction on input image size. Furthermore, based on the universal approximation theorem, as long as the number of neurons in the network is sufficiently large and the connections between neurons are sufficiently complex, it is possible to establish an accurate mapping relationship between input and output. FCNs break through image-level classification to achieve pixel-level classification, making them suitable for the prediction of storage layer parameters. FCNs can automatically extract features from data through training, reducing manual intervention and the need to set initial models during inverse modelling. Compared with other neural networks, FCNs have a relatively simple structure, offer multi-layer perception, and achieve better prediction performance. The following introduces several key structures of fully convolutional networks: convolutional layers, pooling layers, deconvolutional layers, and skip connections.

(1) Convolutional Layer

In a fully convolutional neural network, the convolutional layer is composed of multiple convolutional kernels. The parameters of a convolutional kernel include the kernel size, stride, and padding, which are determined by the backpropagation calculations of the network. These three parameters determine the size of the feature map output by each convolutional layer. Among them, the size of the convolutional kernel can be set arbitrarily to be smaller than the input image size. The size of the convolutional kernel affects the complexity of the input features. The convolutional kernel performs convolution with the feature matrix input from the previous layer, resulting in a feature map of smaller dimensions. The calculation method is as follows:

$$x_j^k = f\left(\sum_{i \in N_j} x_i^{k-1} * w_{ij}^k + b_j^k\right) \tag{6.1}$$

In the equation, $f$ denotes the activation function, $k$ represents the $x_i^{k-1}$th feature matrix of the k − 1 layer, $k-1$ indicates the set of input feature matrices, $i$ denotes the weight parameters of the convolution kernel, and $N_j$ represents the $w_{ij}^k$th bias matrix of the $b_j^k$ layer.

It can be seen that the convolution kernel is essentially a matrix of weighting parameters. By performing convolution and summation between the convolution kernel and the corresponding pixels of the input image, and then applying an activation function, the output image can be obtained. The activation function introduces non-linear mapping into the network. Among these, the ReLU function (Eq. 5.34) is the most commonly used activation function. When the input value is less than 0, the output is 0; when the input value is greater than 0, the output equals the input value. ReLU offers advantages such as high computational efficiency and simplicity, and can effectively avoid the problem of gradient vanishing during network training.

$$f(x) = \max(0, x) \tag{6.2}$$

(2) Pooling Layer

The pooling layer, commonly referred to as the subsampling layer, is used to select and filter the feature maps output by the convolutional layer, removing redundant features from the feature maps, reducing the number of parameters, and preventing overfitting in the network. The selection method in the pooling region is consistent with the scanning approach used in the convolutional layer for feature maps; therefore, the pooling layer is also controlled by three parameters: pooling size, stride, and padding. At present, there are three main pooling methods: max pooling, mean pooling, and stochastic pooling. In this work, the max pooling method is adopted, where the maximum value within a local region is taken as the output of the function, i.e., the local maximum value of the extracted features. The calculation method for max pooling is given by the following formula:

$$u_w = \max_{i,j} R_{i,j}^w \tag{6.3}$$

In the equation, $R_{i,j}^w$ denotes the characteristic map of the kernel $w$ obtained at position $(i, j)$, and $u_w$ represents the calculated local maximum value of the characteristic.

(3) Transposed Convolutional Layer

The transposed convolutional layer, also known as the upsampling layer, represents the primary distinction between fully convolutional neural networks and conventional convolutional neural networks. In traditional convolutional neural networks, a fully connected layer is employed for the final classification processing, whereas in fully convolutional neural networks, the network is constructed using transposed convolutional layers. Transposed convolution is essentially the reverse operation of convolution; its kernel parameters are identical to those of the convolutional layer, but it enlarges the feature map after pooling, effectively performing an interpolation process. The implementation process includes: first, multiplying the input feature map by the transposed convolution kernel, combined with the intermediate matrix and the stride along the horizontal and vertical axes; second, summing the pixel values within the overlapping regions to extract the expanded matrix of the input features; third, adding a bias term to supplement detailed information, and cropping the resulting pixel matrix according to the target image size.

The transposed convolution layer is the core component of fully convolutional neural networks. It enables the mapping of extracted features back to images with the same dimensions as the original input. Consequently, fully convolutional neural networks are not constrained by the input image size.

(4) Skip Connections

After the convolutional layers, pooling layers, and deconvolutional layers, image segmentation can be achieved; however, the accuracy is limited, and some detailed information may be lost. Therefore, skip connections have been introduced into fully convolutional neural networks, which combine the prediction results of the final layer with those of certain preceding layers to complete the network's final prediction. After each pooling operation in the network, the image is downsampled to the original size of $R^{w}_{i,j}$. Assuming the network contains four pooling layers, the image is downsampled to $w$ of its original size. Although the final layer contains most of the information from the input image, direct upsampling at this stage results in the loss of a significant amount of fine detail. By upsampling the results from several preceding layers and adding them to the final feature map, the prediction accuracy of the network can be improved to a certain extent.

The parameters of the reservoir are obtained from seismic observation data through an inversion process, which can be expressed by Eq. (6.4):

$$E(v) = \arg\min L\{f(v), d_{obs}\} \tag{6.4}$$

In the equation, $E$ denotes the objective function, $L$ represents the loss function, with the L2 norm adopted in this study, $v$ refers to the velocity model of the reservoir layer, $f$ indicates the forward modelling process, and $d_{obs}$ denotes the observed data.

Full waveform inversion is considered a highly accurate method for reservoir parameter inversion; however, it is extremely time-consuming. Therefore, deep learning methods can be employed to simulate this process, as shown in Eq. (6.5):

$$out = \psi(in; \Phi) = S(K_2 * (M(R(K_1 * x + b_1))) + b_2) \tag{6.5}$$

In the equation, *in* and *out* represent the input and output data of the network, respectively. ψ indicates that deep learning establishes a nonlinear mapping relationship between the input and output data. $M(\cdot)$ denotes the network parameters of the FCN. $\psi$ is the loss function, while $\Phi$, $S(\cdot)$, $K_1$, and $K_2$ are the convolutional weights. $b_1$ is the downsampling function, for which max-pooling is adopted in this work $b_2R(\cdot)$.

Based on U-Net, a fully convolutional network (FCN) was developed for predicting temporal shift model discrepancies. The network architecture is illustrated in Fig. 6.1. The slices represent feature maps and comprise a contracting path (upper) and an expansive path (lower), each consisting of four steps. Each step contains two identical convolutions, employing 3 × 3 convolutional kernels, ReLU activation functions, and a batch normalisation (BN) layer. In the contracting path, downsampling is performed using max pooling, while in the expansive path, transposed convolutional layers with a kernel size of 2 × 2 are used. In the final layer of the network, a convolutional layer with a kernel size of 1 × 1 is applied. To ensure that the feature map size remains unchanged after convolution, padding mode is also employed.

For the optimisation of the network parameter Φ, the stochastic gradient descent (SGD) method was employed. Given N training data samples, the objective function *in* was defined using Eq. (6.6):

$$E = \frac{1}{N}\sum_{i=1}^{N}(v_i - \psi(in))^2 \tag{6.6}$$

In the equation, $v_i$ represents the actual velocity model, *in* denotes the nonlinear mapping relationship of the FCN network, and $\psi$ is the input data.

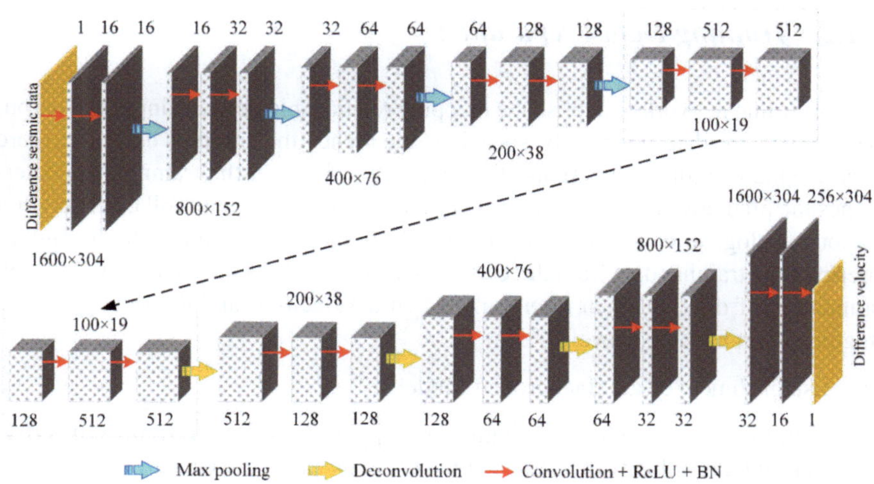


**Fig. 6.1** Architecture of the FCN network

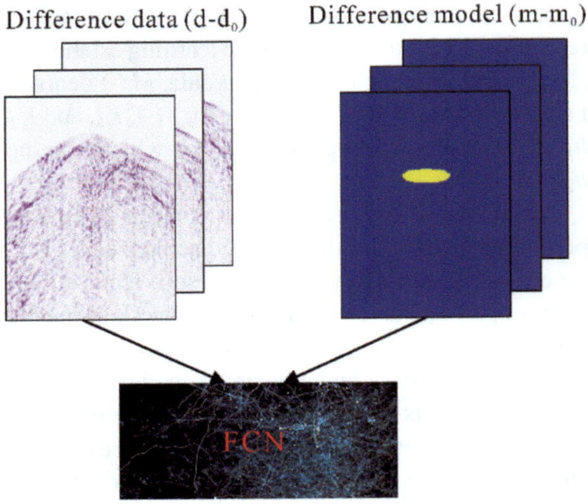


**Fig. 6.2** FCN training method based on time-lapse difference data

The differences in time-lapse seismic data effectively reflect reservoir changes induced by $CO_2$ injection. Therefore, this study proposes a deep learning method based on time-lapse differential data for $CO_2$ storage monitoring. As shown in Fig. 6.2, the time-lapse differential data (monitoring data minus baseline data) and the simulated reservoir velocity changes caused by $CO_2$ (monitoring model minus baseline model) are used as input data to train the FCN network. The trained FCN network incorporates a robust mapping relationship between the differential data and the velocity changes. When new differential monitoring data are input, the local variations in reservoir velocity can be obtained.

### *6.1.2 Training Data Preparation*

Deep learning networks establish the mapping relationship between input and output data driven by data. Generally speaking, the richer the training data, the more accurate the resulting predictions. Therefore, the quality of the training data determines the predictive capability of the neural network. Taking into full consideration various geological conditions, $CO_2$ plume morphology, and data noise, the author prepared the training data from three aspects: construction of the background model, simulation of differences in reservoir time-shift velocities, and the setting of data augmentation and noise levels.

(1) Establishment of the Background Model

Although the FCN establishes a nonlinear mapping between the time-shift difference data and the velocity difference of the reservoir layer, the background model is indispensable for providing the time-depth relationship of seismic data. Moreover, the complexity of the background model affects the complexity of the mapping relationship learned by the neural network. In this book, based on the concept of random functions, for a given total model depth, the thickness and number of geological layers are generated randomly, and the P-wave velocity of each layer is also randomly generated within a specified range, which is set to 2500–4800 m/s. One of the resulting velocity models is shown in Fig. 6.3a, comprising a total of 20 layers. To make the model more consistent with geological conditions, undulating features of the strata are simulated, as illustrated in Fig. 6.3b.

(2) Differences in Temporal Displacement Rates of Storage Layers

After $CO_2$ is injected at the wellhead, the heterogeneity of the geological formation and the development of fractures and other features can result in a variety of possible $CO_2$ plume morphologies. For example, Fig. 6.4a presents a monitoring model following $CO_2$ injection, while Fig. 6.4b illustrates the process of velocity variation and its integration with the background model. At this stage, the velocity value decreases by 100 m/s.

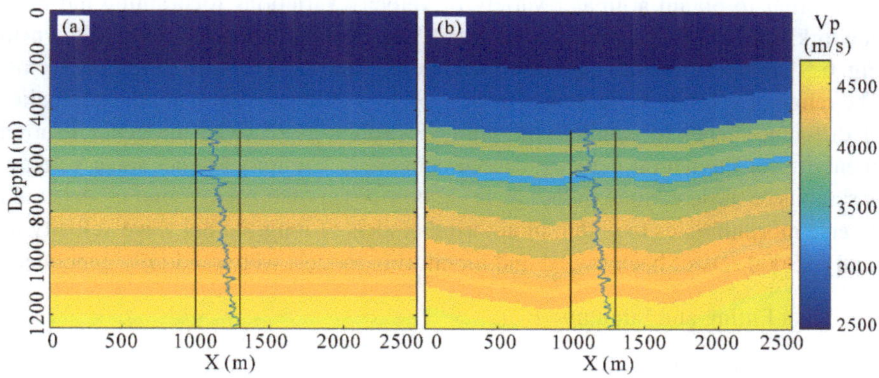


**Fig. 6.3** Randomly generated P-wave velocity background model

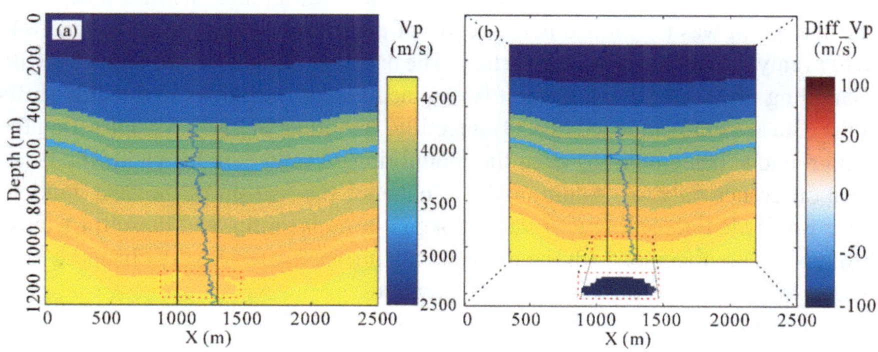


**Fig. 6.4** Monitoring model and velocity changes induced by $CO_2$ injection

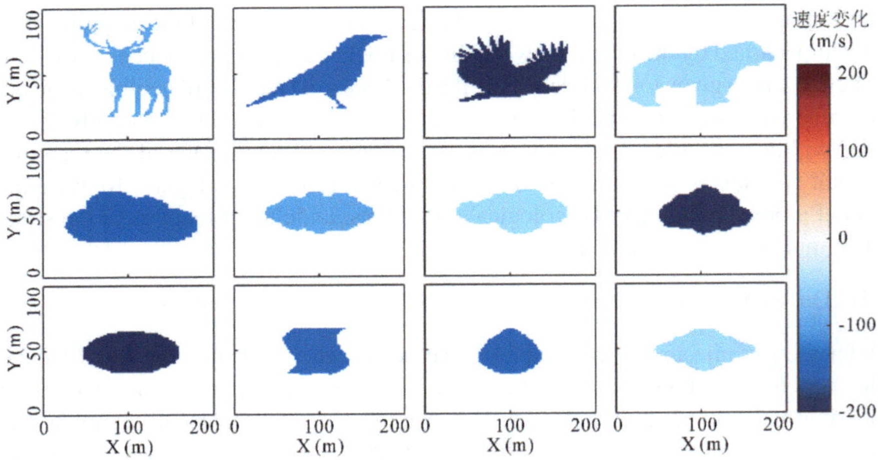


**Fig. 6.5** Simulated patterns of velocity variation

In order to obtain a greater variety of velocity variations for training the FCN network, the author generated a range of shapes, with some examples shown in Fig. 6.5, including various animals, birds, cloud patterns, and basic geometric forms. The range of velocity variations is from − 50 to − 200 m/s. By using the anomalies of different shapes to simulate the plume morphology of $CO_2$, the generalisation capability of the neural network can be enhanced. The longitudinal wave velocity variations of these various shapes were superimposed onto background models with diverse morphologies to obtain monitoring models of longitudinal wave velocity. A total of 3000 sets of background and monitoring models were randomly generated.

(3) Data Fitting and Simulation

A total of 3000 randomly generated models were resampled to a dimension of 256 × 304, and the corresponding shear wave velocities and densities were obtained using the Gardner equation. Seismic data for both the background and monitoring models were generated via time-domain elastic wave finite-difference forward modelling. The shot point was positioned at the centre of the model surface, and 304 receivers were evenly distributed along the surface. The data recording duration was 3.2 s, with a sampling interval of 2 ms. Here, it is assumed that all variations in the time shifts of the data are solely induced by $CO_2$ injection. As shown in Fig. 6.6, subtracting the background seismic data (b) from the monitoring seismic data (a) yields the time-lapse difference data (c). Additionally, to enhance the realism of the data, random noise was added according to a random signal-to-noise ratio, and these time-lapse difference data, together with the corresponding time-lapse velocity changes, were used as input data for training the FCN network.

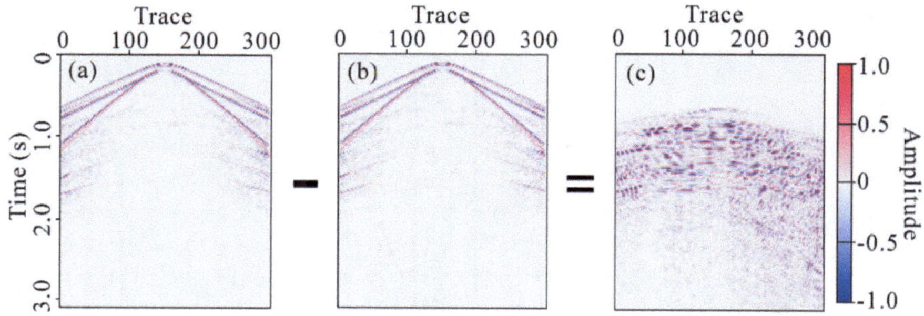


**Fig. 6.6** Method for obtaining time-lapse difference data

**Table 6.1** FCN network training parameters

| Number of iterations | Learning rate | Batch size | Training dataset size | Test dataset size |
|---|---|---|---|---|
| 100 | 0.001 | 10 | 2800 | 200 |
| Model size | Data size | Number of shots | Activation function | Optimizer |
| 256 × 304 | 1600 × 304 | 1 | ReLU | Adam |

### *6.1.3 FCN Network Training*

Based on the 3000 sets of generated data, 2800 sets were used for network training, while the remaining 200 sets were reserved for network testing. Only the intermediate data from one borehole in the model were used during training, and the training data included samples with varying noise levels. The range of time-shift velocities varied from − 50 to − 200 m/s. The training parameters for the FCN network are listed in Table 6.1. After 100 iterations, the training results of the network are shown in Fig. 6.7, which illustrates the network's performance as a function of the number of iterations: (a) shows the loss function value, representing the mean squared error between the actual and predicted velocity variations; (b) shows the prediction accuracy of the network. It can be observed that after approximately 40 iterations, the network establishes a good mapping relationship between the input and output data, achieving relatively high accuracy.

### *6.1.4 Prediction of Temporal Variation in Slip Layer Velocity*

(1) Noise-Free Data

Based on the well-trained FCN network, predictions were tested on some noise-free data (not included in the training dataset). As shown in Fig. 6.8, the velocity varies from − 50 to − 200 m/s, encompassing a range of different shapes. It can be observed that the FCN network proposed by the author, which is based on differential data,

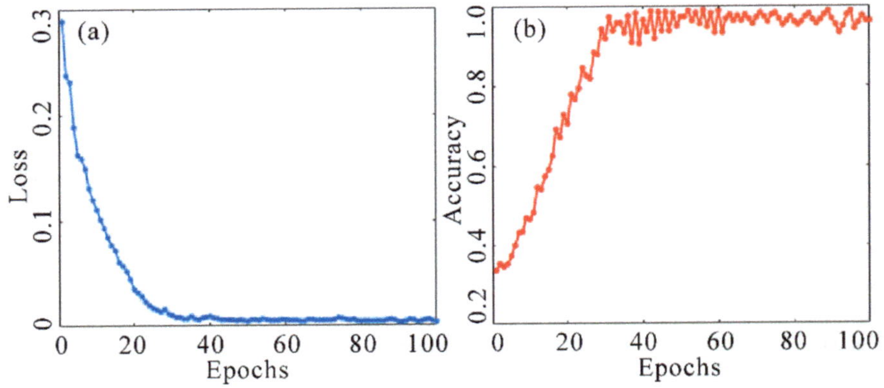


**Fig. 6.7** Variation in FCN network performance with the number of iterations

is able to accurately predict the wake morphology and velocity variations of $CO_2$ under noise-free conditions, demonstrating a high level of predictive accuracy.

(2) Data Containing Noise

The robustness of the trained FCN network to noisy data was tested using datasets containing noise. Figure 6.9 presents a comparison between noiseless and noisy single-shot data, where (a) represents the noiseless data and (b) represents the noisy data (SNR = 15).

The prediction results for noisy data (not included in the training dataset) are shown in Fig. 6.10. It can be observed that the pre-trained FCN network demonstrates good predictive performance when applied to noisy data, accurately predicting both the morphology and velocity variations of the $CO_2$ plume.

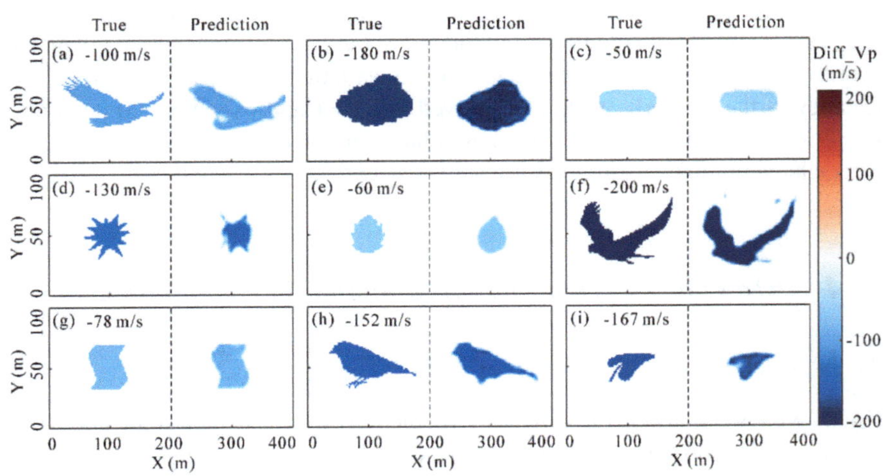


**Fig. 6.8** FCN prediction results for noise-free data

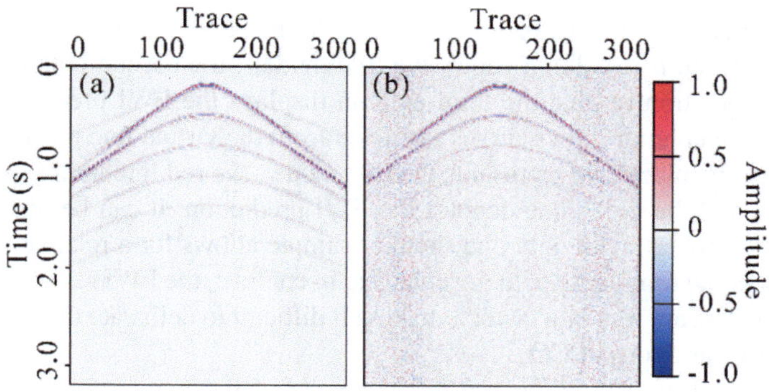


**Fig. 6.9** Noise-free and noisy seismic records

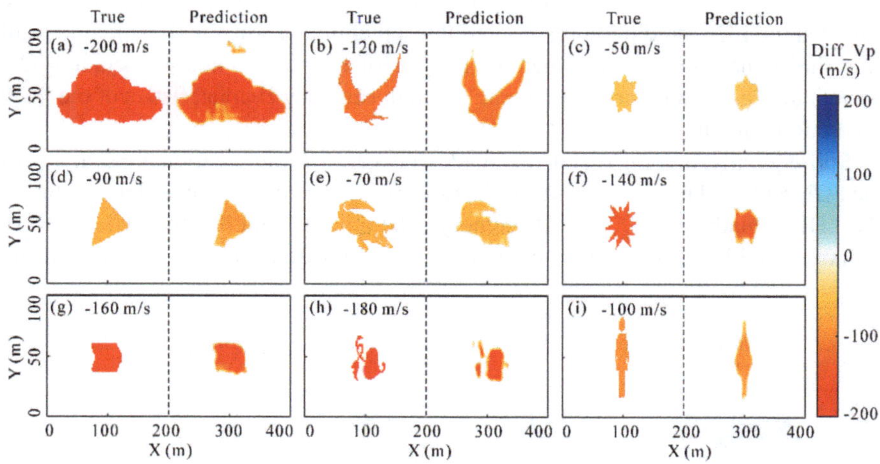


**Fig. 6.10** FCN prediction results for noisy data

The prediction results from both noise-free and noisy data indicate that the FCN network based on time-shift difference data can achieve satisfactory prediction performance even when using data from only a single shot, and it exhibits a certain degree of robustness to noise.

### *6.1.5 Comparison of FCN and FWI Prediction Results*

The author compares the FCN prediction results with those of conventional FWI under noise-free data conditions. The FWI employs a time-domain elastic wave method. As shown in Fig. 6.11, the results are presented for both FCN and FWI using data from only a single shot. Three models are displayed here: (a), (b), and (c), with velocity variations of − 70 m/s, − 120 m/s, and − 180 m/s, respectively. In the figure, the first column shows the true models, the second column presents the FCN prediction results, the third column displays the FWI prediction results, and the fourth column shows curves extracted from the virtual line positions. In the curve plots, the black line represents the true model, the red line indicates the FCN prediction, and the green line denotes the FWI prediction. It can be observed that training the network with only one shot per sample allows for a relatively accurate prediction of the time-lapse velocity changes. In contrast, the FWI results using only single-shot data are less satisfactory, making it difficult to delineate the plume shape and velocity variations of $CO_2$.

To further demonstrate the prediction results of the two methods, the author employed the forward modelling approach to obtain the residuals of the Z-component wavefield. As shown in Fig. 6.12a–c correspond to the wavefield residuals of Fig. 6.11a–c, respectively. Among them, the left side represents the difference between the FCN prediction results and the actual model wavefield, while the right side shows the difference between the FWI prediction results and the actual model wavefield. The results further illustrate the high accuracy characteristics of the FCN prediction outcomes.

Subsequently, a comparison was made between the computational efficiency of the FCN and FWI methods, as shown in Table 6.2. Based on GPU parallel computing (NVIDIA Quadro M5000, 4 GB memory), the training time for the FCN network proposed in this study, which is based on time-shift difference data, was 8.72 h, and

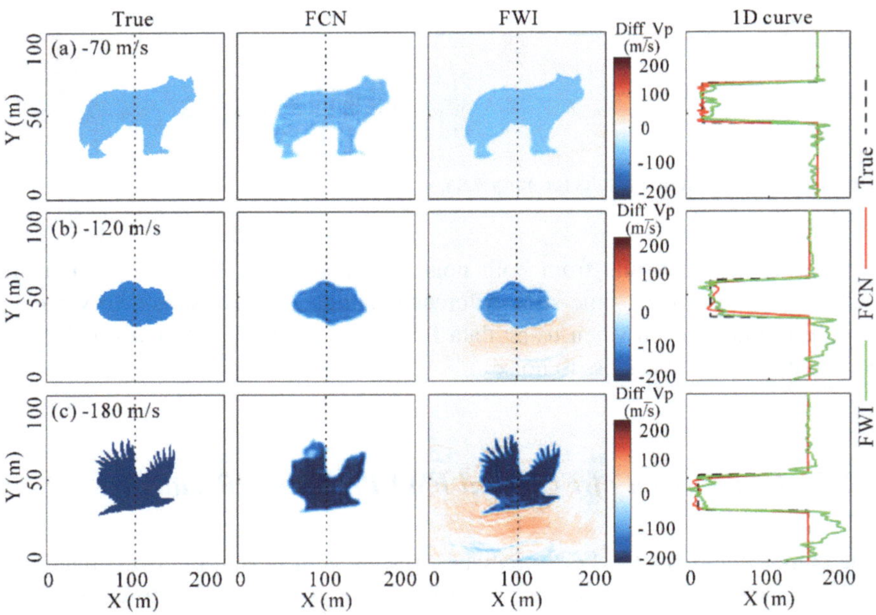


**Fig. 6.11** Comparison of FCN and FWI prediction results for noise-free data

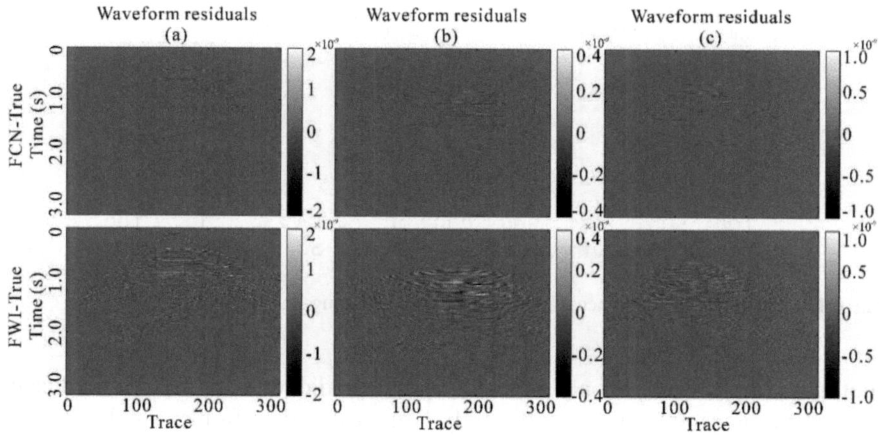


**Fig. 6.12** Z-component data residuals between FCN and FWI prediction results

**Table 6.2** Computational efficiency of FCN and FWI

| Method | Training time | Prediction time | | |
|---|---|---|---|---|
| | | Model (a) | Model (b) | Model (c) |
| FCN | 8.72 h | 3 s | 3 s | 3 s |
| FWI | N/A | 12.2 h | 14.5 h | 13.6 h |

the prediction time for each model was 3 s. In contrast, the computation time for FWI exceeded 12 h. This demonstrates that once the FCN network has been trained, it can rapidly generate predictions without the need for retraining, thereby greatly improving computational efficiency.

### *6.1.6 Analysis of Factors Affecting FCN Prediction Accuracy*

In response to the geological requirements for $CO_2$ storage monitoring, this study analyses the prediction performance of the network under varying training data volumes, different noise levels, and different magnitudes of velocity anomalies. Figure 6.13 illustrates the prediction accuracy of the network with different data volumes (with the number of iterations set to 100): (a) shows the results for different numbers of training data samples (each sample containing data from one shot), and (b) presents the results for varying numbers of shots per sample (with the total number of samples fixed at 2800). As shown in Fig. 6.13a, the prediction accuracy of the network increases with the number of training data samples. When the number of samples reaches 2500, the prediction accuracy attains 97.7%. Further increases in the number of samples yield only marginal improvements in accuracy. Figure 6.13b demonstrates that, when the number of samples is 2800, increasing the number of

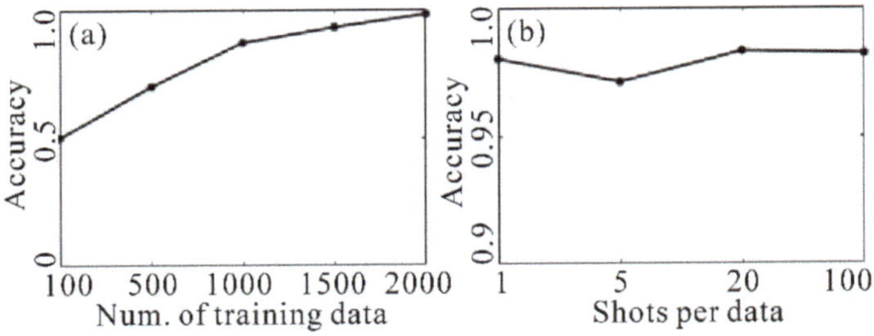


**Fig. 6.13** Network prediction accuracy under different training-data volumes

shots per sample results in only minor changes in prediction accuracy. This indicates that, when the sample size is sufficiently large, using data from a single shot per sample can already achieve high prediction accuracy (approximately 98%).

The 2800 training data sets used encompass data at various noise levels, which enhances the network's noise resistance. To demonstrate this conclusion, the author trained the FCN network using two different approaches and tested the prediction accuracy of both networks on data with different noise levels. Network 1 was trained exclusively on noise-free data, while Network 2 was trained, as previously described, using data at different noise levels. The quantity and parameters of the training data for Network 1 and Network 2 were identical. As shown in Fig. 6.14a, the prediction accuracy of the two networks varies with noise level: the black line represents the prediction accuracy of Network 1, and the red line that of Network 2. The test results indicate that the prediction accuracy of both networks increases with the signal-to-noise ratio. However, Network 2 consistently demonstrates higher accuracy overall, indicating that training the network with data at multiple noise levels is beneficial for the prediction of noisy data.

The velocity variation induced after $CO_2$ injection is correlated with the degree of $CO_2$ saturation; that is, the greater the $CO_2$ saturation, the larger the velocity change. In order to test the range of velocity variations that the FCN network can predict, the author used data with velocity changes ranging from − 10 to − 300 m/s for network prediction. Figure 6.14b shows the prediction accuracy of the network under different velocity variations. The green line represents noise-free data, while the blue

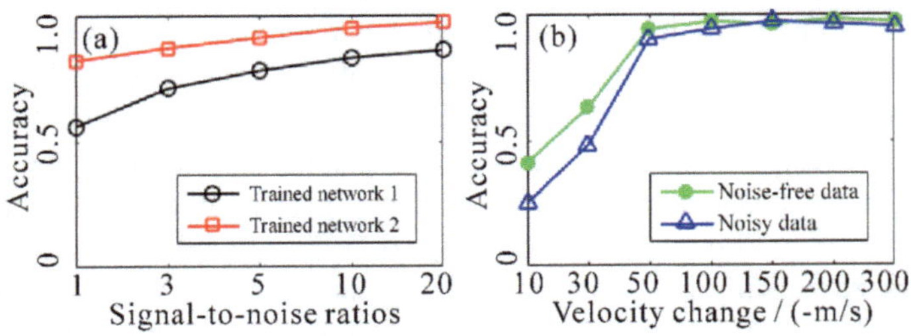


**Fig. 6.14** Network prediction accuracy under different noise levels and magnitudes of velocity variation

line represents data containing noise. It can be observed that the network achieves relatively high prediction accuracy when the velocity variation exceeds 50 m/s. When the velocity is less than 50 m/s, due to the minimal data differences, the predictive capability is insufficient. Therefore, the FCN network based on time-shift difference data proposed by the author can achieve accurate prediction of velocity variations greater than 50 m/s under both theoretical data with and without noise. Compared with the FWI inversion results described in the previous section, this method demonstrates better performance in predicting reservoir parameter variations and is suitable for geological monitoring of $CO_2$ storage and migration.

## 6.2 Carbon Dioxide Migration Monitoring Technology Based on Material Parameters

In time-lapse seismic exploration, variations in the amplitude of seismic reflected waves typically represent the combined response of multiple subsurface parameters, making it difficult to distinguish the contribution of each parameter. This poses a significant challenge for high-resolution prediction of subsurface parameters. Deep learning possesses strong nonlinear processing capabilities; therefore, this book integrates deep learning to develop methods for the simultaneous prediction of multiple subsurface parameters. Using $CO_2$ geological storage as an example, the changes in subsurface parameters during the $CO_2$ injection process are simulated, enabling $CO_2$ migration monitoring based on subsurface parameter estimation.

The workflow of the multi-task learning-based reservoir multi-parameter prediction method for $CO_2$ geological storage monitoring can be described as shown in Fig. 6.15, comprising two stages. Stage 1 involves obtaining the training dataset from basic geological parameters and production parameters, which includes three components: numerical simulation, petrophysical modelling, and seismic forward modelling. Firstly, numerical simulation is used to obtain the variations in reservoir parameters at different periods of $CO_2$ injection. Secondly, petrophysical models are employed to calculate the elastic parameters of the strata (including P-wave velocity, S-wave velocity, and density). Finally, seismic forward modelling is conducted to generate angle-gather seismic data. Stage 2 consists of the training and prediction processes of the neural network. Based on the sufficiently rich training data obtained in Stage 1, the network training and prediction tasks are completed.

### *6.2.1 Carbon Dioxide Migration Numerical Simulation and Shale Physical Model*

Numerical simulation is an effective method for understanding the migration of $CO_2$ and the variation of reservoir property parameters under specific geological and

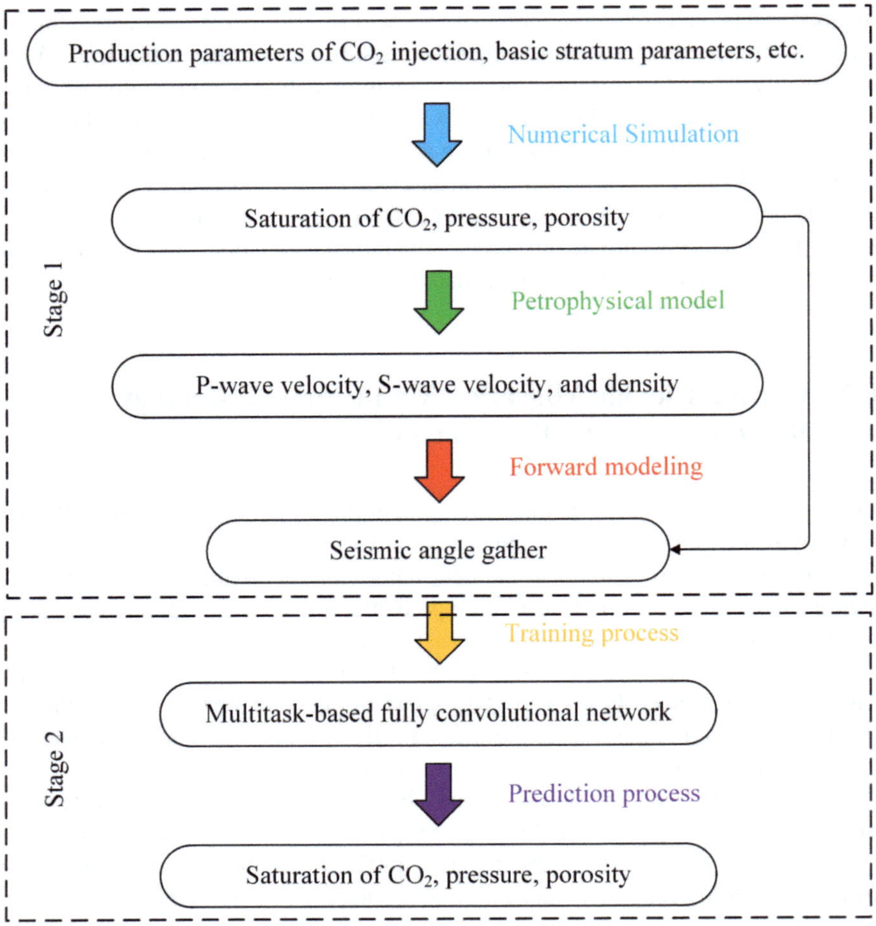


**Fig. 6.15** Workflow of the multi-parameter reservoir prediction method based on multi-task learning

operational conditions. In this study, numerical simulation is employed to obtain the changes in reservoir parameters corresponding to different stages of $CO_2$ injection, thereby providing sufficient data for the training of deep learning networks.

In $CO_2$-driven oil recovery engineering, seepage exhibits multiphase and multi-component characteristics, conforming to Darcy's law. The seepage velocity α of a single flow phase $v_\alpha$ can be expressed as:

$$v_\alpha = \frac{KK_{r\alpha}}{\mu_\alpha}(\nabla P_\alpha + \rho_\alpha g \nabla Z) \tag{6.7}$$

In the equation, $K$ denotes the permeability of the porous medium, $\nabla P$ is the relative permeability of the fluid within the medium, $\rho_\alpha$ represents the fluid viscosity,

$\nabla P$ is the pressure gradient, $g$ denotes the fluid density, $K_{r\alpha}$ is the gravitational acceleration, and $\mu_\alpha$ indicates the vertical coordinate.

The flow of $CO_2$ in porous media is governed by the mass and energy conservation equations, which can be expressed as follows:

$$\frac{\partial}{\partial t}\left[\phi\sum_{\alpha}\left(x_{\alpha}^{i}\rho_{\alpha}s_{\alpha}\right)\right]+\nabla\cdot\sum_{\alpha}\left(x_{\alpha}^{i}\rho_{\alpha}v_{\alpha}-\phi s_{\alpha}\tau_{\alpha}D_{\alpha}\rho_{\alpha}\nabla x_{\alpha}^{i}\right)-S_{i}=0 \tag{6.8}$$

$$\frac{\partial}{\partial t}\left[\phi\sum_{\alpha}(U_{\alpha}\rho_{\alpha}s_{\alpha})+(1-\phi)\rho_{r}C_{r}T\right] +\sum_{\alpha}\nabla\cdot(\rho_{\alpha}H_{\alpha}v_{\alpha})-\nabla\cdot(\lambda\nabla T)+\varepsilon\sigma T^{4}-S_{H}=0 \tag{6.9}$$

In the equation, $T$ denotes porosity, $\phi$ represents mass fraction, $x$ indicates the component of a certain phase α, $H$ denotes saturation, $\lambda$ represents curvature, $i$ denotes dispersion coefficient, $\alpha$ represents mass source, $s_\alpha$ denotes specific internal energy, $\tau$ represents specific heat capacity, $D$ denotes temperature, $S_i$ represents specific enthalpy, $\varepsilon$ denotes electrical conductivity, $U$ represents radiation coefficient, $\sigma$ denotes the Stefan–Boltzmann constant, and $C$ represents heat quantity.

Based on the main equations above, combined with the initial and boundary conditions, the finite difference method can be employed to solve the partial differential equations. A more detailed description will not be provided here.

After $CO_2$ is injected through the injection well, fluid displacement occurs within the reservoir, which in turn leads to changes in the reservoir's elastic parameters. Firstly, $CO_2$ injection alters the saturation of different fluids within the pore spaces. Secondly, the injection of high-pressure $CO_2$ induces changes in the effective stress of the reservoir. Thirdly, as a pressurised fracturing solvent, $CO_2$ can modify the connectivity of the reservoir, thereby causing variations in porosity.

The traditional Gassmann equation does not take into account the influence of pressure variations on elastic parameters, and therefore does not meet the simulation requirements for $CO_2$ storage. Consequently, we employ the Hertz-Mindlin model to establish the relationship between pressure, porosity, and the dry rock bulk modulus and shear modulus. The pressure-dependent dry rock bulk modulus $\phi$ proposed by Hertz can be expressed as:

$$K_{dry}=\sqrt[3]{\frac{C^{2}(1-\phi)^{2}\mu_{m}^{2}P_{e}}{18\pi^{2}(1-\nu)^{2}}} \tag{6.10}$$

The shear modulus of dry rock under variable confining pressure, as proposed by Mindlin, $\mu_{dry}$

$$\mu_{dry}=\frac{5-4\nu}{5(2-\nu)}\sqrt[3]{\frac{3C^{2}(1-\phi)^{2}\mu_{m}^{2}P_{e}}{2\pi^{2}(1-\nu)^{2}}} \tag{6.11}$$

In the formula, $\mu_m$ denotes porosity, $\mu_m$ represents the compaction ratio, $\phi$ is the effective stress, and $C$ is the shear modulus of the sandstone framework. Based on the mineral composition and bulk modulus of the sandstone, the V-R-H method can be used to calculate μm. $v$ is the coordination number, which is generally taken as a constant. In this work, the relationship between $P_e$ $C$

$$C = W * \left(11.759 * e^{1-\phi} - 12.748\right) \tag{6.12}$$

Based on the obtained $K_{dry}$ and $\mu_{dry}$, and in conjunction with the Gassmann equations (formulas 6.13 and 6.14), the P-wave and S-wave velocities as well as the density after $CO_2$ saturation can be calculated.

$$K_{sat} = K_{dry} + \frac{\left(1 - \frac{K_{dry}}{K_m}\right)^2}{\frac{\phi}{K_{fl}} + \frac{1-\phi}{K_m} - \frac{K_{dry}}{K_m^2}} \tag{6.13}$$

$$\mu_{sat} = \mu_{dry} \tag{6.14}$$

In the formula, $K_{sat}$, $K_{dry}$, $K_m$, and $K_{fl}$ represent the bulk moduli of saturated rock, dry rock, rock matrix, and pore fluid, respectively; $\mu_{sat}$ and $\mu_{dry}$ are the shear moduli of saturated rock and dry rock, respectively; $\mu_{dry}$ denotes porosity. In this study, the pores are assumed to be isolated, spherical, and the variation in porosity does not alter the reservoir $\phi$

$$V_p' = \sqrt{\frac{M_{sat}'}{\rho_b'}} \tag{6.15}$$

$$V_s' = \sqrt{\frac{\mu_{sat}'}{\rho_b'}} \tag{6.16}$$

$$\rho_b' = (1 - \phi)\rho_m + \phi\rho_{fl}' \tag{6.17}$$

In general, after $CO_2$ flooding, the longitudinal and transverse wave velocities of the reservoir are jointly controlled by porosity $\phi$, fluid saturation $S$, effective geostatic pressure $P_e$, and the fundamental parameters of the rock matrix $Rockpara$, and can be expressed as $V_p', V_s' \rightarrow (\phi, S, P_e, Rock\_para)$. Among these, $Rockpara$ represents the intrinsic parameters of the rock, which can be directly obtained from relevant reservoir data and remain unchanged during the $CO_2$ injection process. Therefore, during $CO_2$ flooding, the variation in the elastic parameters of the reservoir is mainly influenced by the three parameters: porosity $\phi$, fluid saturation $S$, and effective geostatic pressure $P_e$. Furthermore, it can be concluded that the variation in time-lapse seismic amplitude is primarily caused by changes in $\phi$, $S$ and $P_e$.

A simple layered curve can be used to illustrate the $CO_2$ displacement process in $CO_2$-EOR. As shown in Fig. 6.16, the variations in P-wave velocity, S-wave velocity, and density with respect to $CO_2$ saturation, effective stress, and porosity are simulated. Panels (a), (d), and (g) show the changes in P-wave velocity (Vp), where it can be observed that $CO_2$ saturation, effective stress ($P_e$), and porosity all influence the P-wave velocity. The velocity variation can reach up to 300 m/s. The effect of $CO_2$ saturation is relatively minor; only when transitioning from gas-free to gas-bearing conditions does the P-wave velocity decrease significantly, after which further increases in gas saturation have little impact on the P-wave velocity. Panels (b), (e), and (h) display the changes in S-wave velocity (Vs), which exhibit a similar trend to that of the P-wave velocity, with variations reaching up to 200 m/s. Panels (c), (f), and (i) show the changes in density (Den), indicating that density is unaffected by effective stress, is most influenced by porosity, and can vary by up to 0.1 g/cm$^3$.

To clearly illustrate the variation trends of P-wave and S-wave velocities as well as density, numerical data at a depth of 600 m were extracted. The initial state of the formation was set with a $CO_2$ saturation of 0%, an effective stress of 30 MPa, and a porosity of 0.28. In the simulation, when one parameter was varied, the others remained constant. As shown in Fig. 6.17, it can be observed that when the $CO_2$

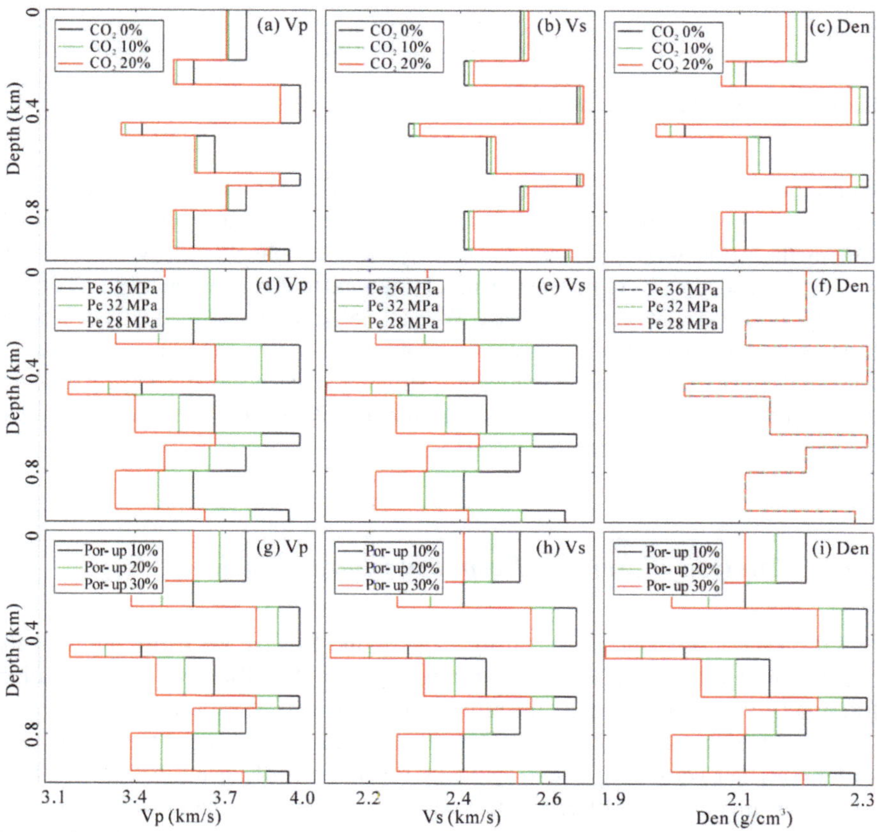


**Fig. 6.16** Simulated variations in formation elastic parameters after $CO_2$ injection

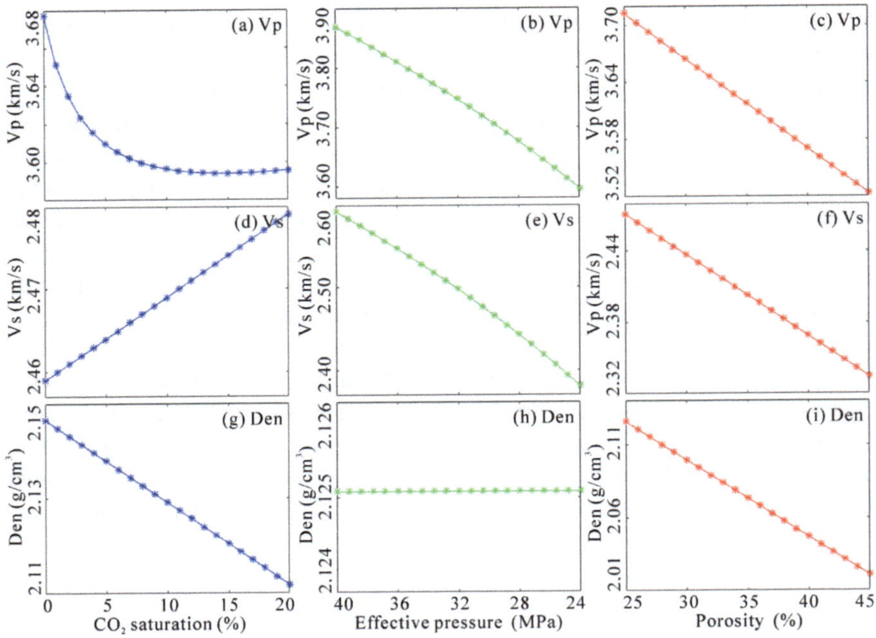


**Fig. 6.17** Trend of formation elastic parameter variation at 600 m

saturation changes from 0 to 20%, with increasing $CO_2$ saturation, both the P-wave velocity and density of the formation decrease, while the S-wave velocity increases. Among these, the S-wave velocity and density exhibit linear changes, whereas the P-wave velocity decreases rapidly at the initial stage and then remains nearly unchanged as $CO_2$ saturation continues to increase. When the effective stress gradually decreases from 40 to 24 MPa (effective stress = total stress − pore pressure), both the P-wave and S-wave velocities decrease approximately linearly, but the density remains unchanged. When the porosity increases from 25 to 45%, the P-wave velocity, S-wave velocity, and density all decrease linearly.

### 6.2.2 *Analysis of Seismic Response Characteristics*

Based on the aforementioned data for P-wave and S-wave velocities and density under different conditions, the author employed the Aki–Richards approximation formula (Eq. 6.23) to calculate the reflection coefficients at various incident angles (offsets) (Aki and Richards 1980). The reflection coefficients were then convolved with the Ricker wavelet to obtain synthetic seismic records as a function of offset. As shown in Fig. 6.18, fifteen reflection gathers were simulated at 2° intervals, covering angles from 0 to 30°. The first row displays the reflection angle gathers for the initial reservoir state and for cases where $CO_2$ saturation, effective pressure, and porosity are varied individually. The second row shows the corresponding differences relative to the initial state. In the initial reservoir state, the $CO_2$ saturation is 0%, the effective pressure is 30 MPa, and the porosity varies from 0.2 to 0.35 depending on the formation. The simulations are all based on the initial reservoir state, with only one parameter altered at a time. From the boxes and arrows in the difference plots, it can be seen that changes in $CO_2$ saturation, effective pressure, and porosity all induce variations in seismic reflection amplitude, and the amplitude differences manifest differently across various angle ranges. According to the current reservoir parameters, changes in $CO_2$ saturation have a greater impact at small angles, changes in effective pressure have a more pronounced effect at large angles, and variations in porosity exhibit significant differences at different angles. Therefore, the author uses angle gather data at different angles as the basis for predicting reservoir $CO_2$ saturation, effective pressure, and porosity parameters, which is beneficial for feature extraction by neural networks.

$$R(\theta) = a\frac{\Delta V_P}{V_P} + b\frac{\Delta\rho}{\rho} + c\frac{\Delta V_S}{V_S} \tag{6.18}$$

In the equation, $c = -4\left(\frac{V_S}{V_P}\right)^2 \sin^2\theta$, $V_P$ represents the longitudinal wave velocity. $a = \frac{1}{2\cos^2\theta}$, $\theta$ denotes the angle. $b = 0.5 - \left[2\left(\frac{V_S}{V_P}\right)^2 \sin^2\theta\right]$, $V_S$ is the transverse wave velocity, and ρ indicates the density. $\rho = \frac{\rho_2+\rho_1}{2}$, $\Delta\rho = \rho_2 - \rho_1$, $V_P = \frac{V_{P2}+V_{P1}}{2}$, $\Delta V_P = V_{P2} - V_{P1}$, $V_S = \frac{V_{S2}+V_{S1}}{2}$, $\Delta V_S = V_{S2} - V_{S1}$, $\theta = \frac{\theta_i+\theta_t}{2}$.

Based on the above analysis, it can be concluded that, in $CO_2$ geological storage, variations in $CO_2$ saturation, effective stress, and porosity all have an impact on seismic amplitude. In other words, the differences in time-lapse seismic responses are the result of the combined effects of changes in $CO_2$ saturation, effective stress, and porosity. Moreover, these differences are related to offset distance, with different parameter variations being sensitive within different offset distance ranges. Therefore, this study uses angle gather data within various angular ranges as the basis for predicting multiple parameter changes during $CO_2$ geological storage.

### *6.2.3 Fully Convolutional Neural Network Architecture Based on Multi-task Learning*

In recent years, multi-task learning networks have achieved notable success in the processing and interpretation of seismic data. Based on a fully convolutional neural network, the present study establishes a multi-task learning network for the prediction of multiple geological parameters, as illustrated in Fig. 6.19. The network comprises one contracting path (encoder) and three expanding paths (decoders), each dedicated to the prediction of a different parameter. At each step, two identical convolutional

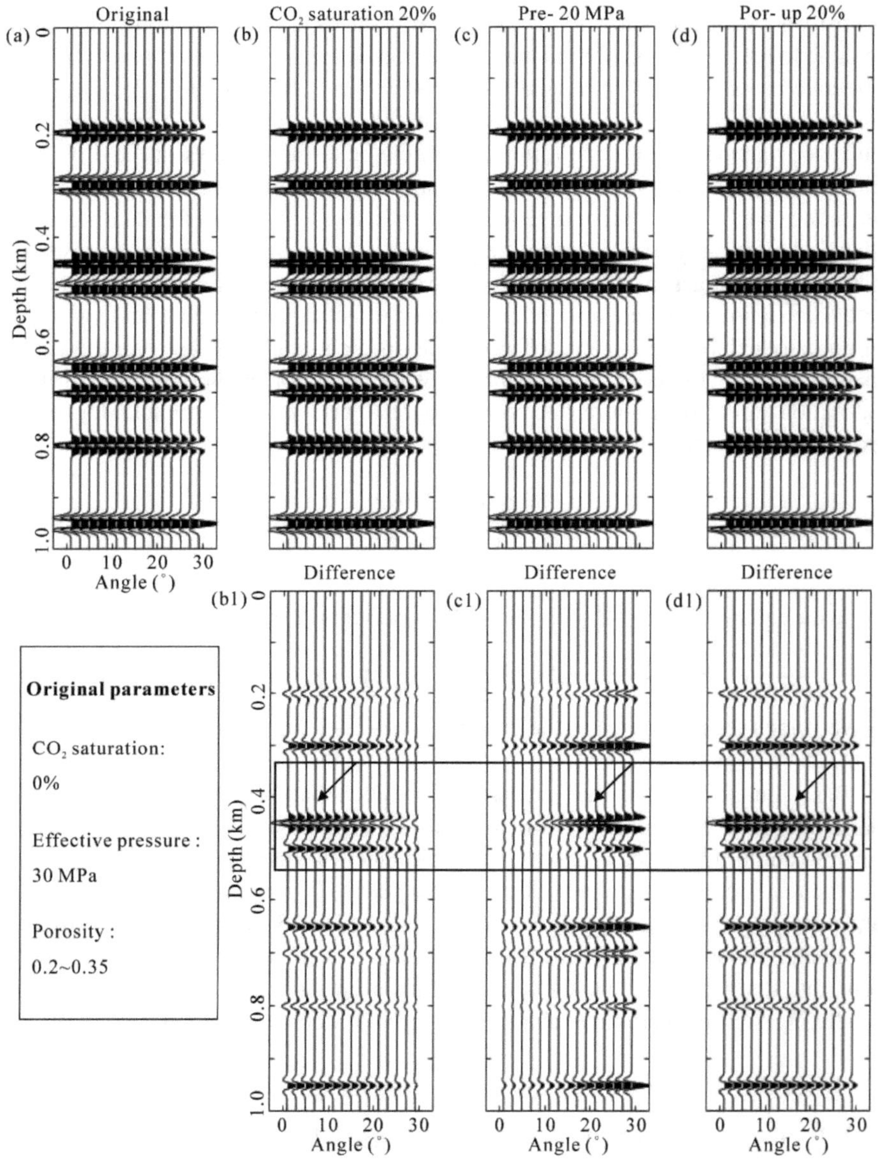


**Fig. 6.18** Seismic reflection differences

processes are performed, indicated by red arrows, utilising 3 × 3 convolution kernels, ReLU activation, and batch normalisation (BN). Padding mode is adopted in each convolutional operation, and the kernel size of the convolution in the transposed convolutional network is 2 × 2. The network input consists of seismic gather data at various angles, while the outputs are three parameters: $CO_2$ saturation, porosity, and pore pressure. The three tasks share the network's features, thereby enhancing prediction accuracy.

The prediction of reservoir parameters based on deep learning is a typical regression problem. In this study, the mean squared error is used to separately calculate the loss functions for different tasks, which are then combined to form the total loss function of the network, $a = \frac{1}{2\cos^2\theta}$, as shown in Eq. (6.19):

$$\begin{aligned} L_{total} &= \lambda_1 \cdot L_{sat} + \lambda_2 \cdot L_{por} + \lambda_3 \cdot L_{pre} \\ &= \frac{1}{N}\left(\lambda_1 \sum_{i=1}^{N} (S_i - S_i')^2 + \lambda_2 \sum_{i=1}^{N} (Por_i - Por_i')^2 + \lambda_3 \sum_{i=1}^{N} (\mathrm{Pre}_i - \mathrm{Pre}_i')^2\right) \end{aligned} \tag{6.19}$$

In the equation, $L_{sat}$, $L_{por}$, and $L_{pre}$ represent the loss functions for the prediction of $CO_2$ saturation, porosity, and pressure, respectively. λ1, λ2, and λ3 denote the weighting factors. Since the numerical ranges of $CO_2$ saturation and porosity are both 0–1, while the pressure values are in the tens, in this study, we select $\lambda 1 = 1$, $\lambda 2 = 1$, and $\lambda 3 = 0.05$. $\lambda_1$ denotes the total number of pixels, and $\lambda_2$, $\lambda_3$, and $\lambda_1 = 1$ represent the $CO_2$ saturation, porosity, and pressure, respectively, while $'$ denotes the predicted value.

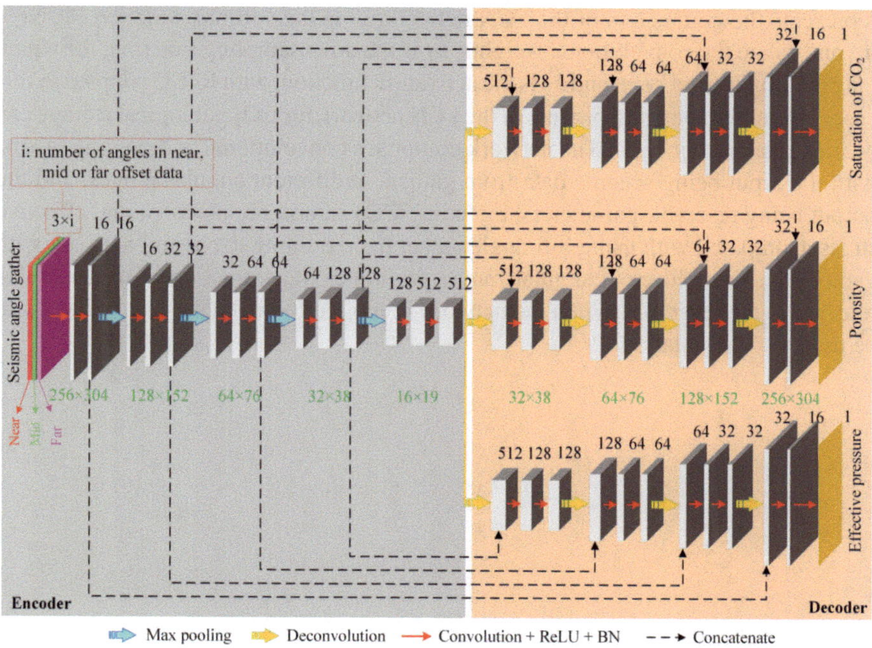


**Fig. 6.19** Architecture of the multi-task learning network

### 6.2.4 Simultaneous Prediction Process for Multiple Geological Layer Parameters

Obtaining subsurface parameters from seismic data is an inversion process, which can be described as follows:

$$E(m) = \arg\min L\{f(m), d_{obs}\} \tag{6.20}$$

In the equation, $E$ denotes the misfit function, $L$ represents the loss function, m refers to the set of subsurface parameter data, $f$ indicates the synthetic modelling process, and $d_{obs}$ denotes the observed seismic data.

Seismic inversion is typically ill-posed, characterised by uncertainty and non-uniqueness. Based on the strong nonlinear processing capability of deep learning, the author employs a fully convolutional neural network (FCN) to address the seismic inversion problem. The computation of the FCN network can be expressed in the following form:

$$Y = \psi(X; \Phi) = S(a_2 * (P(R(a_1 * x + b_1))) + b_2) \tag{6.21}$$

In the formula, $X$ and $Y$ represent the input and output of the network, respectively. $P(\cdot)$ denotes the nonlinear mapping relationship between the input and output, $R(\cdot)$ represents the parameters of the FCN network, $\psi$ is the softmax function, $\Phi$, $S(\cdot)$, $a_1$, and $a_2$ are the convolutional weights, $b_1$ is the downsampling function, for which max pooling is used here, and $b_2$ is the activation function, with ReLU adopted in this case. The implementation process of the FCN network for $CO_2$ geological storage can be illustrated in Fig. 6.20. The network comprises convolutional and pooling layers, with the input being seismic data from gathers at different angular ranges, and the output being reservoir parameters, including $CO_2$ saturation, porosity, and effective stress. Compared with using full-angle gather data, training the neural network with gather data from different angular ranges facilitates the extraction of data features relevant to the reservoir. When the network is applied to field data, Algorithm 1 can be used for computation.

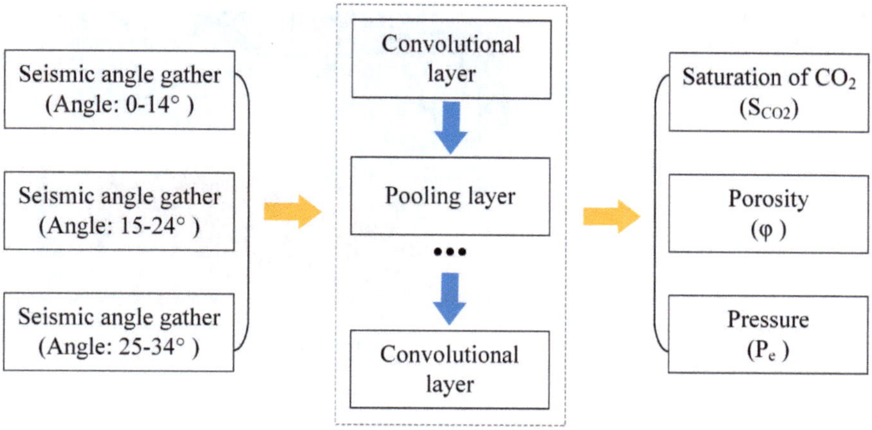


**Fig. 6.20** Implementation workflow of the FCN network for $CO_2$ geological storage

---

Algorithm 1 Simultaneous multi-parameter prediction of formation parameters based on multi-task learning

---

**Training procedure**

Input: seismic angle-gather data with different angles, formation parameters, including $CO_2$ saturation, pore pressure, and porosity, number of iterations, data size, batch size, learning rate, number of training datasets $\{d\}_{n=1}^{N}\{P\}_{n=1}^{N} IterdsbnlrN$

Initialization: $loss = 0y^0 = \{d\}$

for to do $i = 1Iter$
for to do $n = 1N/bn$
$y = S(a_2 * (P(R(a_1 * x + b_1))) + b_2)$;
$loss = L_2(\widetilde{y_n}, y_n)$;
end
$\Theta_{n+1} = Adam(\Theta_n; loss; lr)$; // network parameters $\Theta$
end

**Prediction procedure**

Input: observed time-lapse seismic angle-gather data
$Y = \psi(X; \Phi)$; // trained network $\psi$
Output: formation parameters, including $CO_2$ saturation, pore pressure, and porosity

---

## 6.2.5 *Training Data Preparation and Network Training*

In order to obtain a sufficient number of training samples, a random method was employed to generate geological models with folding structures, including random numbers of strata, layer thicknesses, and folding configurations. As shown in Fig. 6.21, a randomly generated stratigraphic structure is presented, consisting of

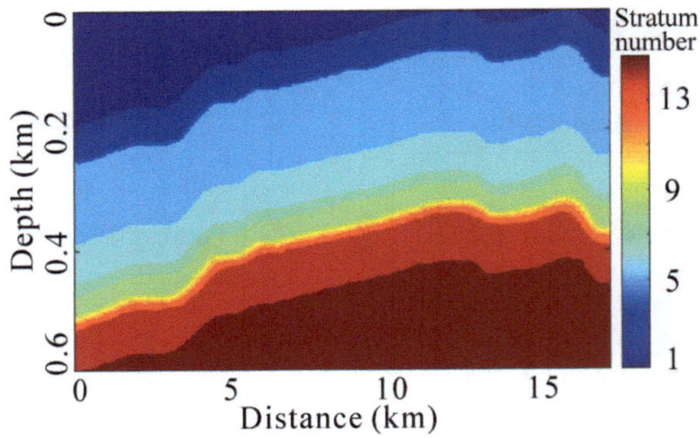


**Fig. 6.21** Simulated stratigraphic structure

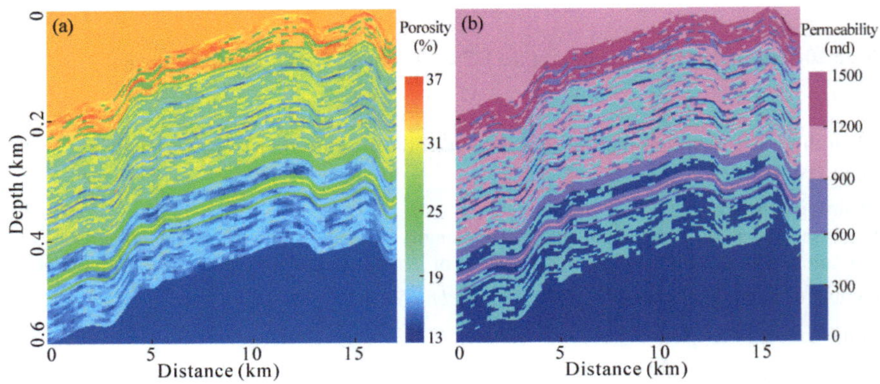


**Fig. 6.22** Random assignment of stratigraphic parameters within a specified range

15 layers with varying folding structures, closely resembling the characteristics of actual geological strata.

Based on the generative stratigraphic structural model, lithological parameters within a certain range are randomly assigned, including porosity, permeability, and effective stress, among others. In order to enhance the realism of the model, different parameters are assigned not only to different strata, but also to different locations within the same stratum, thereby introducing lateral heterogeneity. As shown in Fig. 6.22, the distribution of porosity and permeability parameters exhibits pronounced lateral heterogeneity, reflecting authentic geological characteristics.

By randomly assigning oil reservoir production parameters within a reasonable range, such as $CO_2$ injection rate and oil–gas production rate, and combining these with numerical simulation methods for $CO_2$ storage, it is possible to obtain the migration status of $CO_2$ and the dynamic changes in the reservoir over different periods. For example, Fig. 6.23 illustrates the initial reservoir parameters and those after $CO_2$ injection, with the $CO_2$ injection well located at X = 6200 m. The reservoir state prior to $CO_2$ injection is shown in Fig. 6.23a, c, e. After 25 years of $CO_2$ injection (at an injection rate of 1.3 million tonnes per year), the reservoir parameters are

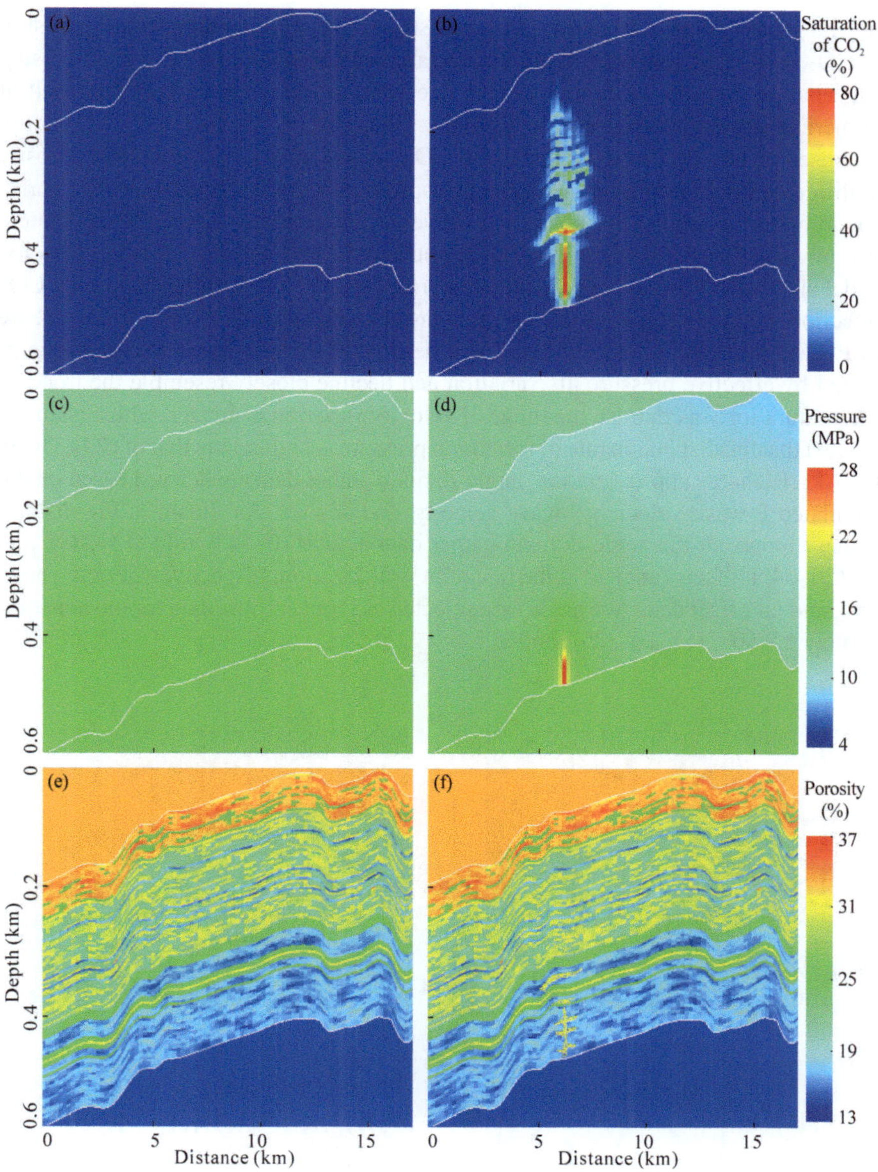


**Fig. 6.23** Numerical simulation results of reservoir parameters before and after $CO_2$ injection

depicted in Fig. 6.23b, d, f, representing $CO_2$ saturation, effective pressure, and porosity, respectively. As seen in (b), the $CO_2$ plume migrates laterally through the reservoir, centred on the well, with $CO_2$ saturation reaching up to 0.8. Figure 6.23d shows that with continuous $CO_2$ injection, the effective pressure around the well increases, while at locations farther from the well, effective pressure decreases due to oil and gas production. Figure 6.23f demonstrates the variation in reservoir porosity; as $CO_2$ is injected, reservoir fractures increase, leading to an overall rise in porosity. The changes in porosity include human intervention and are not solely the result of numerical simulation outcomes.

Based on the simulated parameters of $CO_2$ saturation, pore pressure, and porosity at different times, combined with the petrophysical model of sandstone, the corresponding formation P-wave velocity, S-wave velocity, and density can be calculated. As shown in Fig. 6.24a–c represent the calculated P-wave velocity, S-wave velocity, and density after 25 years of $CO_2$ injection, respectively. The anomalies near the $CO_2$ injection well are clearly visible. An increase in $CO_2$ saturation leads to a decrease in P-wave velocity, S-wave velocity, and density. Since density is essentially unaffected by effective pressure, its variation and decline closely resemble the trend of $CO_2$ saturation. In contrast, the range of velocity variation is greater, which is due to the fact that the dispersion rate of formation pressure is higher than that of $CO_2$. Using the Aki–Richards approximation, angle-domain gather data are derived based on the simulated P-wave velocity, S-wave velocity, and density. As shown in Fig. 6.24d–f, these represent the angle-domain gather data at near (0–14°), mid (15–24°), and far (25–34°) offsets (angles) at the injection well location, respectively. For the near, mid, and far offset data, two angles are selected as input seismic data for the network, namely 5°, 10°, 15°, 20°, 25°, and 30°.

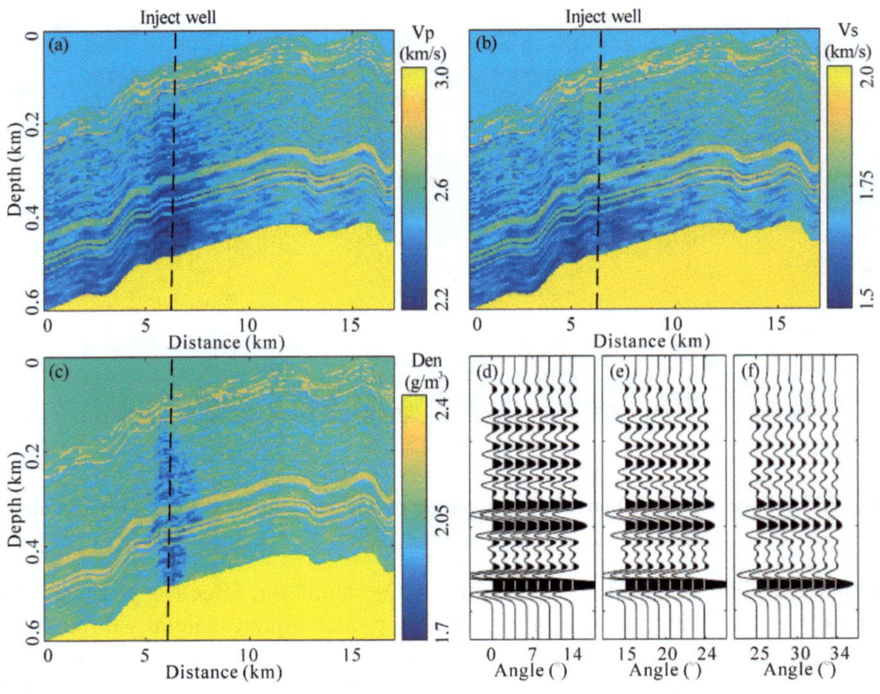


**Fig. 6.24** Model elastic parameters and simulated angle-gather seismic data

In summary, the process of generating network training data in deep learning-based multi-parameter inversion methods mainly comprises the following six steps:

(1) Random generation of initial geological layers.
(2) Randomly assign stratum parameters within a certain range, including porosity, permeability, etc.
(3) Randomly set the production parameters for $CO_2$ flooding.
(4) The target formation parameters at different storage durations were obtained through numerical simulation.
(5) According to the rock physical model, the corresponding longitudinal and transverse wave velocities and density at each time point were obtained.
(6) Obtain sets of ground motion data with different drift distances (angles) through forward modelling simulations.

Through the above steps, a large volume of reservoir parameters exhibiting $CO_2$ migration characteristics and corresponding seismic data were obtained (4700 sets of training data and 300 sets of test data), providing ample training data for the neural network.

For the geological task of multi-parameter prediction of reservoir layers, the network parameters adopted by the author are shown in Table 6.3. The training performance of the network is illustrated in Fig. 6.25, where the validation accuracy of the network is evaluated using the $R^2$ coefficient, calculated according to Eq. (6.22).

$$R^2\_score = 1 - \frac{\sum_i (y_i - \hat{y}_i)^2}{\sum_i \left(y_i - \overline{y}_i\right)^2} \tag{6.22}$$

In the equation, $\overline{y}_i$ denotes the mean value, while $\hat{y}_i$ represents the regression fitting value. It can be observed that the loss value continuously decreases with the increase in the number of iterations. When the number of iterations reaches 30, the loss value falls below 0.01. The validation accuracy of the network increases steadily with the number of iterations; after 30 iterations, the accuracy remains almost unchanged and stays above 0.95. When the number of iterations reaches 100, the accuracy attains 0.99. The training time of this network under GPU parallel computing is 15.6 h (NVIDIA Quadro M5000 with 4G memory).

**Table 6.3** Network parameters for simultaneous multi-parameter prediction

| Activation function | ReLU | Model size | 256 × 304 |
|---|---|---|---|
| Optimizer | Adam | Data size | 256 × 304 × 6 |
| Downsampling function | Max-pooling | Batch size | 64 |
| Convolution kernel | 3 × 3 | Number of iterations | 100 |
| Number of channels | 3 × i = 6 | Learning rate | 0.001 |

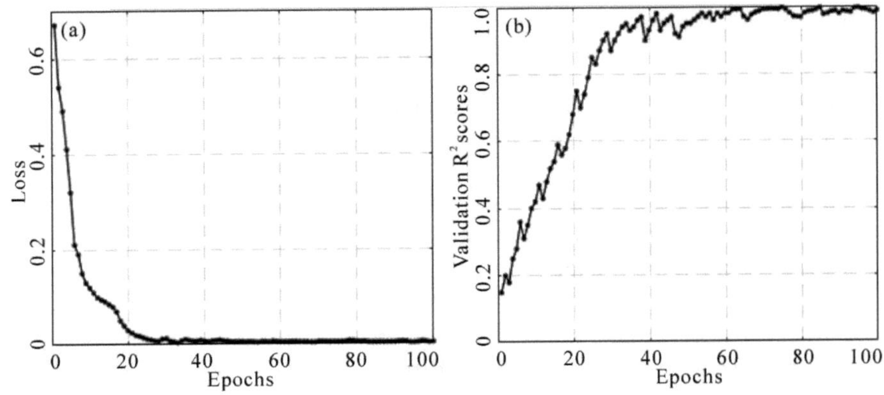


**Fig. 6.25** Network performance

### 6.2.6 *Synthetic Data Testing and Analysis*

The Sleipner CCS project has made available its geological model data and 4D seismic data; this is a saline aquifer $CO_2$ storage project. Firstly, based on a specific 2D cross-sectional dataset from the actual Sleipner geological model, the predictive performance of the algorithm was tested. As shown in Fig. 6.26a presents a cross-section of the Sleipner geological model, (b) shows the simulated $CO_2$ migration after the injection of 12 Mt, while (c) and (d) display the simulated pore pressure and porosity, respectively. Here, the porosity of the formation remains unchanged before and after $CO_2$ injection. From Fig. 6.26b, c, it can be observed that the diffusion rate of pressure is much greater than the migration rate of $CO_2$. Seismic angle gather data were obtained through forward modelling. To make the data more realistic, random Gaussian noise (SNR = 15 dB) was added, calculated as follows:

$$SNR = 10\lg\frac{d_{signal}}{d_{noise}}(\mathrm{dB}) \tag{6.23}$$

As shown in Fig. 6.27, the grain boundary distribution data corresponding to different positions on the cross-section in Fig. 6.26 are presented. In order to evaluate the effectiveness of the neural network, the grain boundary distribution has been divided into three sections: 0–14°, 15–24°, and 25–34°. Six angles—5°, 10°, 15°, 20°, 25°, and 30°—were selected for neural network testing.

The network prediction results are shown in Fig. 6.28, encompassing $CO_2$ saturation, pressure, and porosity. The left side depicts the actual model, while the right side presents the predicted results. It can be observed that, overall, the predicted results exhibit a high degree of agreement with the actual model. Similarly, the $R^2$ coefficient was used to calculate the accuracy of the predictions, which is indicated in the upper left corner of each figure. Figure 6.28b displays the predicted results for $CO_2$ saturation, where the migration contours of $CO_2$ are clearly delineated.

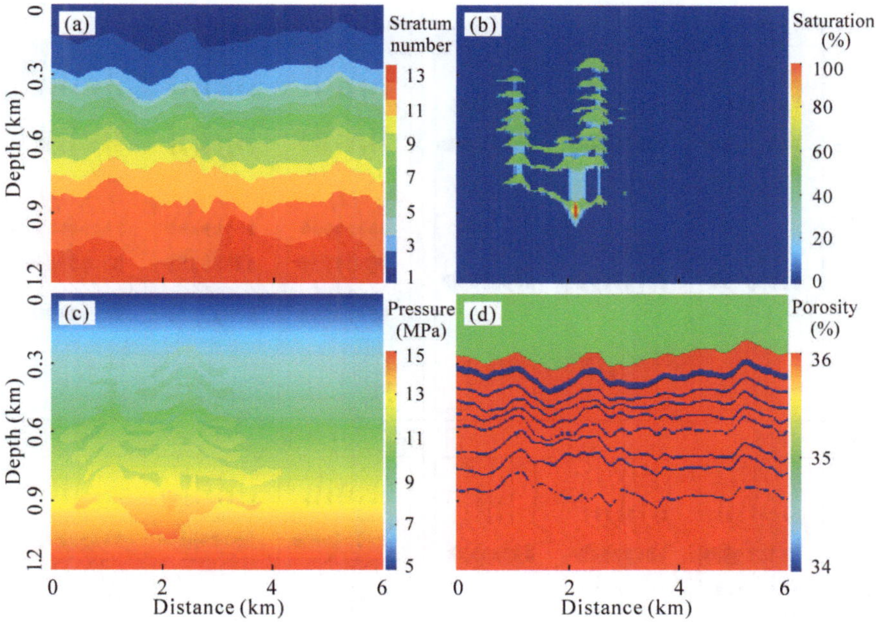


**Fig. 6.26** A stratigraphic section and reservoir parameters for a selected Sleipner interval

Apart from some minor details, the remaining areas correspond well with the actual model (Fig. 6.28a), achieving a prediction accuracy of 0.93. Figure 6.28d shows the predicted results for reservoir pressure. In the non-pressurised variation region, the predicted results reflect a uniform variation with depth, and the range and shape of the pressure variation region correspond closely to the actual model (Fig. 6.28c), with an overall prediction accuracy of 0.94. Figure 6.28f presents the predicted results for reservoir porosity, where the depiction of clay layers with low porosity is clearly defined. Except for the folded structures at the left and right boundaries, the results correspond well with the actual model (Fig. 6.28e), achieving a prediction accuracy of 0.91.

To clearly demonstrate the prediction results, 2570 data points were sampled and a 1D curve was plotted. As shown in Fig. 6.29, the red line represents the predicted results, while the black line corresponds to the actual model. Figure 6.29a displays the predicted results for $CO_2$ saturation. Differences are observed only in the region between 300 and 400 m, while the rest of the curve shows good agreement. These discrepancies occur at the leading edge of $CO_2$ migration. Due to the relatively small scale and the limited seismic resolution, such differences arise. Figure 6.29b presents the predicted pressure results, where both the overall trend and local variations align well with the actual data. Figure 6.29c shows the predicted porosity results, which overall match the actual trend very well, although there are many local variations in the predicted results. This is because the porosity in the network's training data exhibits lateral heterogeneity (see Fig. 6.23e, f), whereas the porosity model for

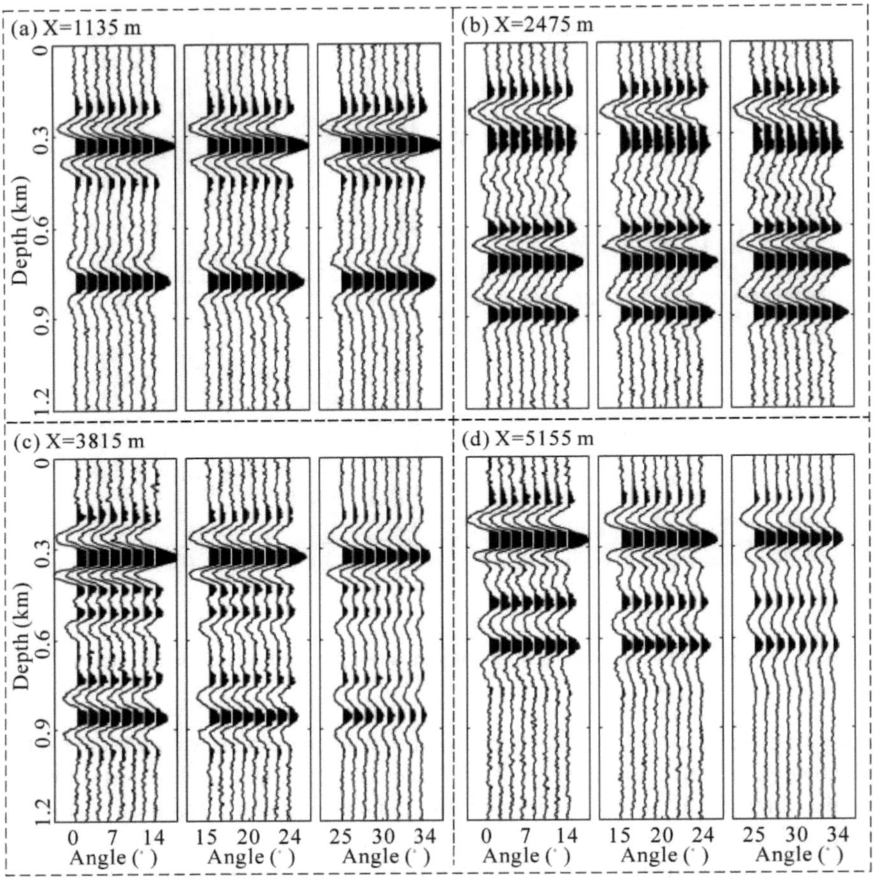


**Fig. 6.27** Simulated seismic angle-gather data for Sleipner

Sleipner assumes uniform porosity within the same geological layer. Therefore, the predicted results display local variations; however, this does not affect the overall predictive performance.

## *6.2.7 Practical Data Application*

From the above prediction results, it can be observed that the multi-parameter simultaneous prediction method proposed by the author achieves satisfactory performance on synthetic datasets with noise. To further evaluate the network's performance, its predictive capability was tested using real data. However, the publicly available seismic data from the Sleipner project only include stacked data at near, mid, and far

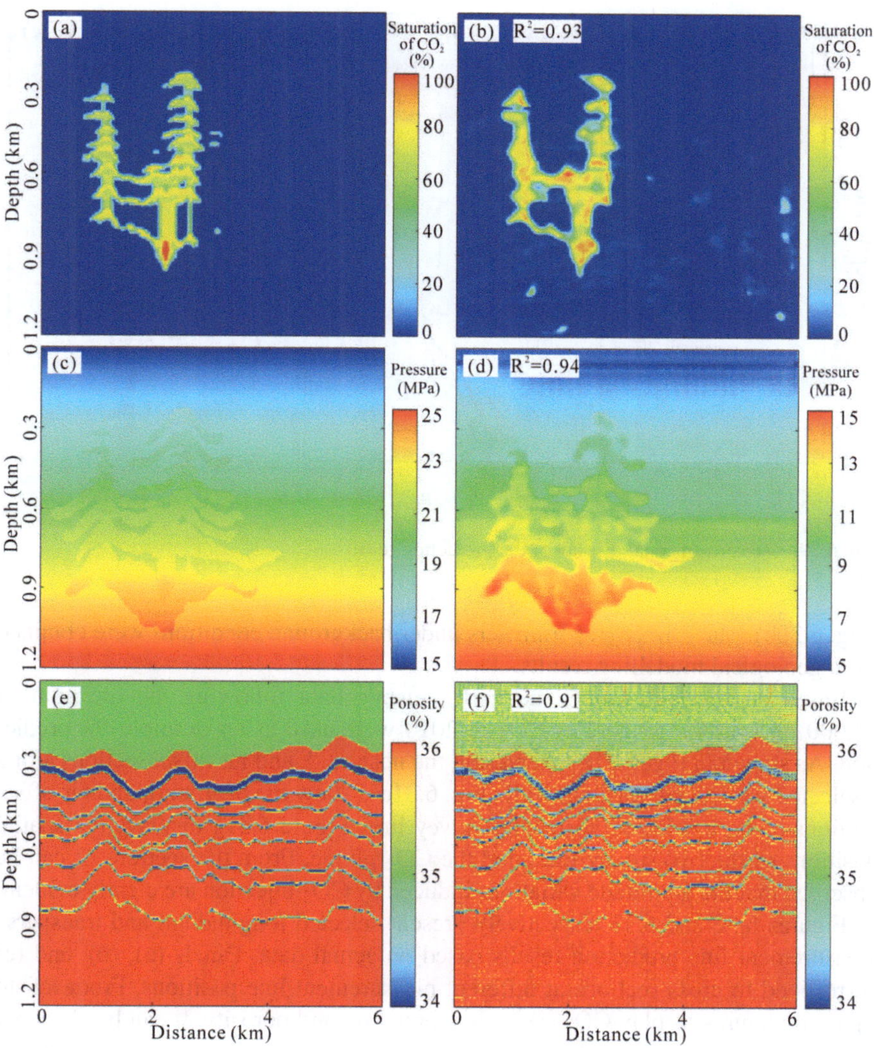


**Fig. 6.28** Prediction results for synthetic data based on the method proposed in this study

offsets, and do not provide pre-stack angle gather data. Therefore, based on the pre-trained network, additional training was conducted using stacked data at different offsets (near, mid, and far), with the training data still derived from the 5000 sets of synthetic models (Fig. 6.30a). Since the pre-trained network is designed for six channels, the near, mid, and far offset data were duplicated to form six channels. In addition, to inform the network of the geological conditions under background circumstances (i.e., prior to $CO_2$ injection), the actual background data and corresponding formation parameters were included in the network training as training data

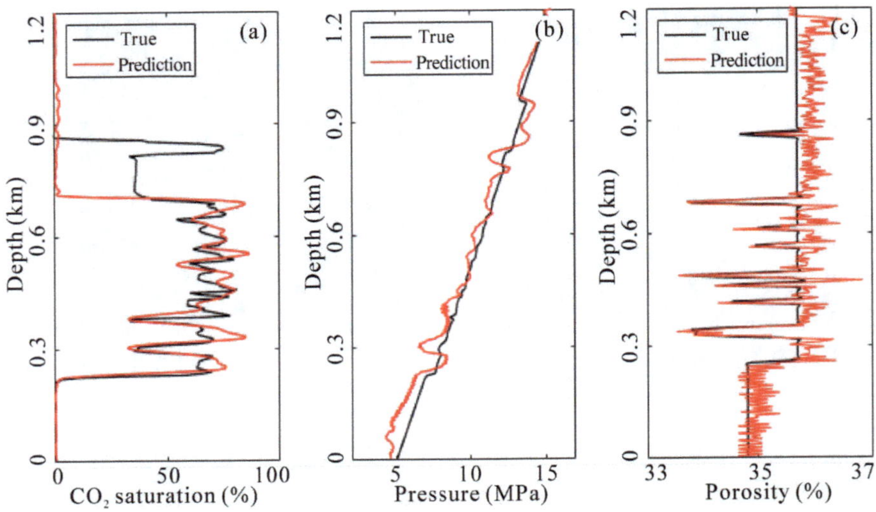


**Fig. 6.29** One-dimensional display of the prediction results

(Fig. 6.30b). The formation parameters under background conditions were obtained from geological modelling results.

Based on the well-trained network, seismic data following the injection of 12,000 mt (million tonnes) of $CO_2$ in 2010 were used as an example for prediction. As shown in Fig. 6.31a–c represent the near, mid, and far offset sections along a selected vertical survey line, while Fig. 6.31d–f show the near, mid, and far offset sections along a selected horizontal survey line. Figure 6.31g displays the spatial positions of the two survey lines and the $CO_2$ plume. From the sections, it can be observed that the amplitude variations induced by $CO_2$ injection are clearly evident.

Figures 6.32 and 6.33 respectively present selected longitudinal and transverse measurement line prediction results based on actual data. Panels (a), (b), and (c) correspond to cross-sections at different measurement line positions. From left to right, the figures display $CO_2$ saturation, pressure, and porosity. It can be observed that the $CO_2$ plume is clearly delineated, local pressure variations are pronounced, and several thin interbedded mudstone layers within the Utsira reservoir are distinctly visible.

To verify the accuracy of the predicted results, a comparison was made between the predicted $CO_2$ saturation on two cross-sections and the results of manual interpretation. As shown in Fig. 6.34, the left and right panels correspond to the cross-sections at the positions of the vertical measurement line (Fig. 6.32b) and the horizontal measurement line (Fig. 6.33b), respectively. The first row presents the results of manual interpretation. Due to the unknown plume thickness in each layer, it was assumed that the entire sand body layer within the plume boundary was fully saturated with $CO_2$. The second row shows the results predicted by the proposed method. It can be observed that the morphology of the $CO_2$ plume predicted by this method aligns well with the manual interpretation. Meanwhile, the calculated SSIM values

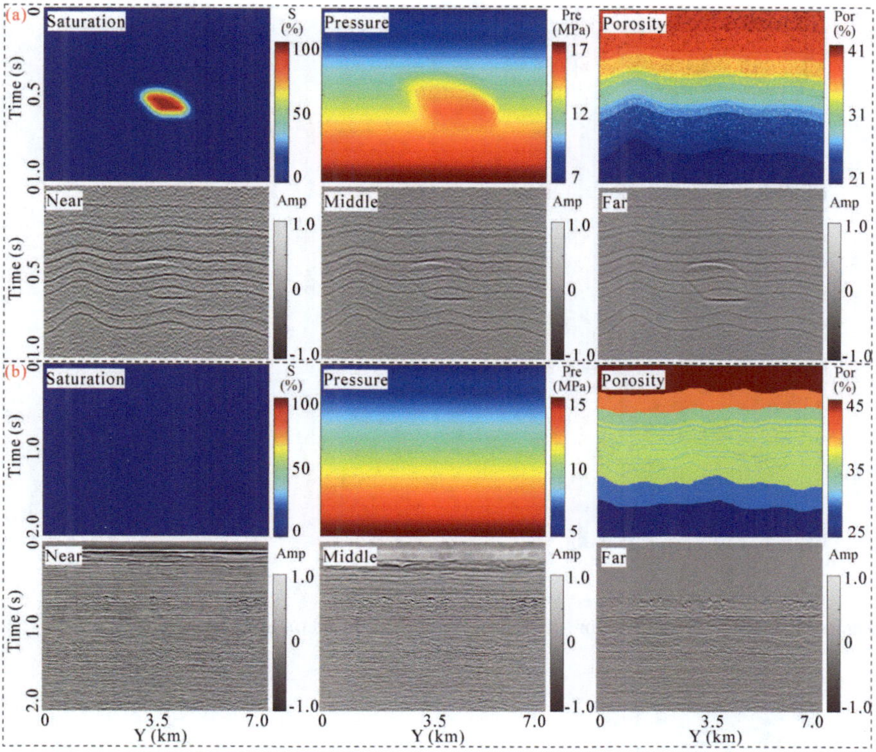


Fig. 6.30 Data and labels for retraining the network

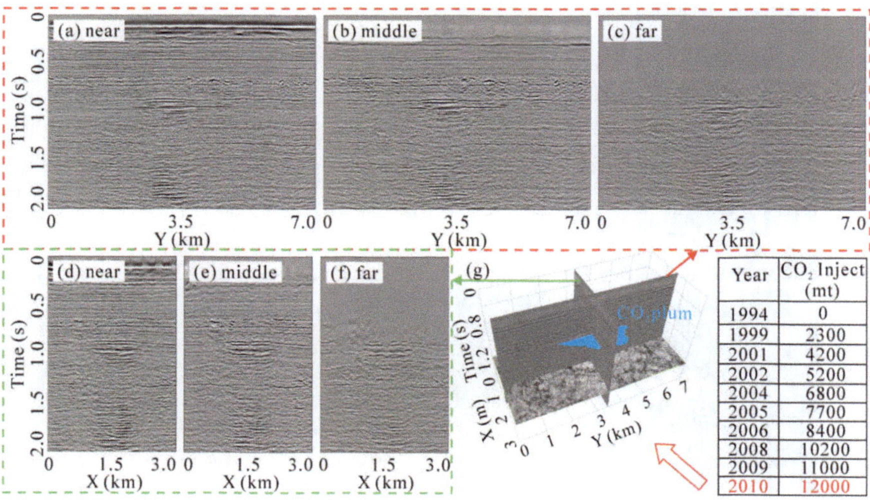


| Year | $CO_2$ Inject (mt) |
|---|---|
| 1994 | 0 |
| 1999 | 2300 |
| 2001 | 4200 |
| 2002 | 5200 |
| 2004 | 6800 |
| 2005 | 7700 |
| 2006 | 8400 |
| 2008 | 10200 |
| 2009 | 11000 |
| 2010 | 12000 |

Fig. 6.31 Actual three-dimensional seismic profile from Sleipner

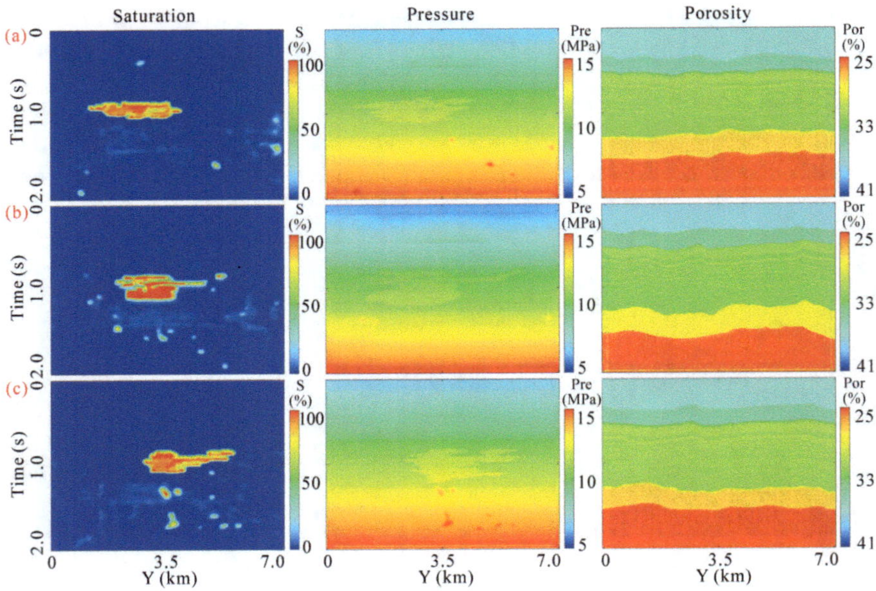


Fig. 6.32 Prediction sections for selected inline profiles based on Sleipner field data

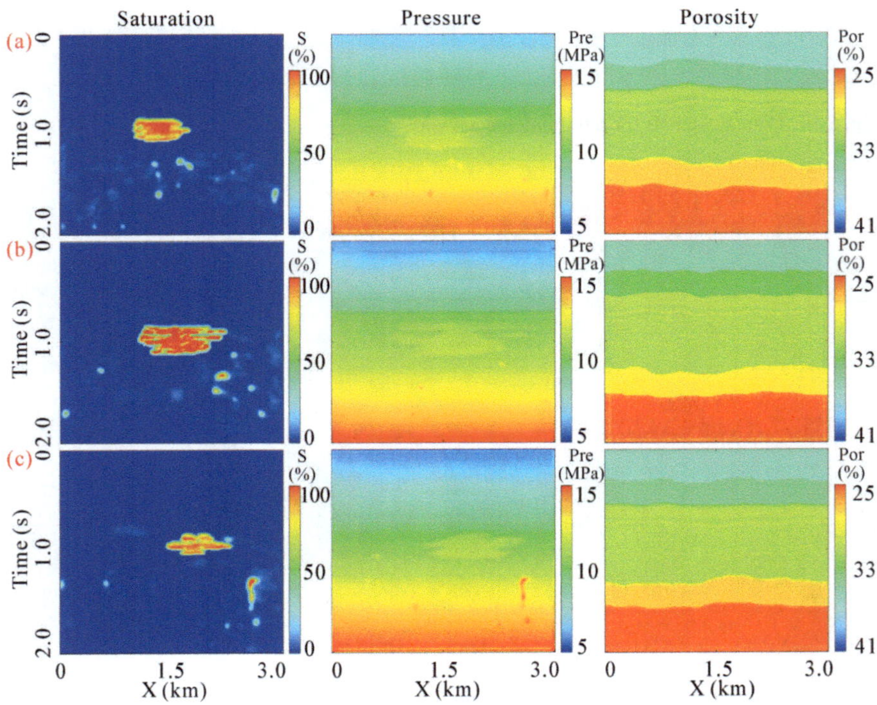


Fig. 6.33 Prediction sections for selected crossline profiles based on Sleipner field data

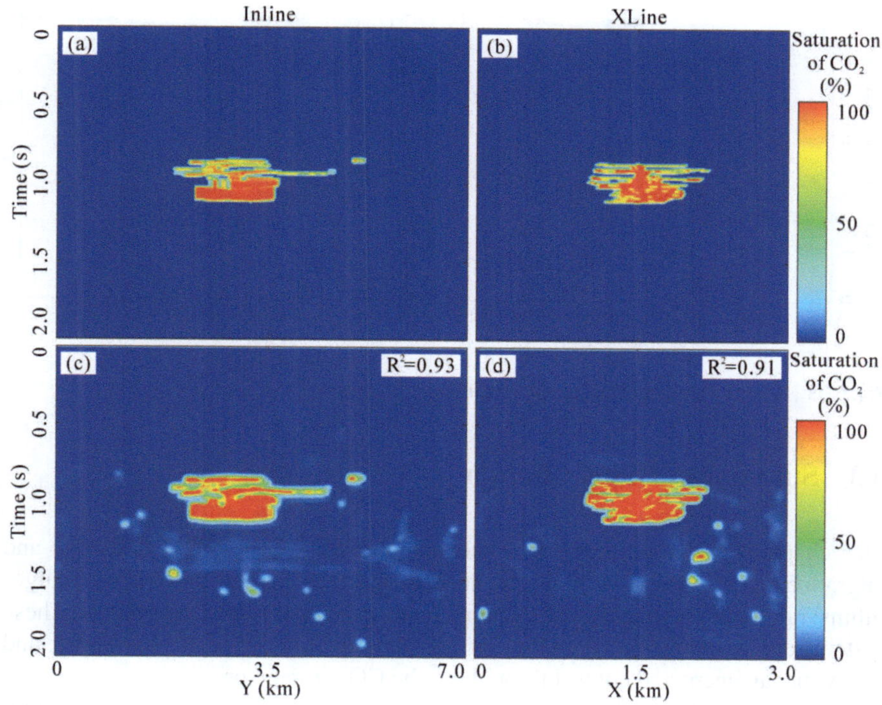


**Fig. 6.34** Comparison between predicted $CO_2$ saturation and manual interpretation

are displayed in the upper right corners of figures (c) and (d), indicating the accuracy of the predicted results. The field test data demonstrate that the multi-parameter prediction network proposed in this study possesses a certain degree of generalisation capability and can achieve satisfactory results in practical applications.

Building upon the foundation of this study, a further analysis was conducted to investigate the effects of the number of training data samples and the signal-to-noise ratio (SNR) of the data on the prediction accuracy of the network. As shown in Fig. 6.35, it can be observed from (a) that the greater the number of training data samples, the higher the prediction accuracy of the network. When the data volume reaches 3000 samples, the prediction accuracy of the network attains 0.9. With further increases in data volume, the improvement in prediction accuracy becomes more gradual. From (b), it can be seen that the higher the SNR of the data, the higher the prediction accuracy of the network. When the SNR exceeds 10 dB, the variation in prediction accuracy is relatively small and remains above 0.92. In the figure, the green arrow indicates the data conditions adopted in this study, namely 4700 training data samples and test data with SNR = 15.

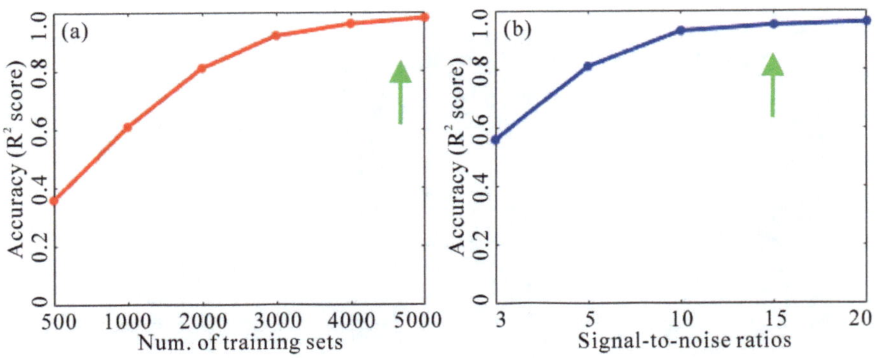


**Fig. 6.35** Network prediction accuracy for different datasets

## 6.3 Summary of This Chapter

This chapter focuses on monitoring technologies for $CO_2$ geological storage and migration, establishing two major technology systems centred on intelligent algorithms: rapid plume imaging and dynamic monitoring of physical parameters. These systems are key supports for achieving real-time tracking, risk early warning, and safety management and control throughout the $CO_2$ storage process.

In response to the demand for rapid imaging of $CO_2$ plume migration, this chapter adopts the fully convolutional neural network (FCN) as the core technology, optimising processes such as training data preparation and network training to enable fast prediction of velocity variations in the time-lapse storage layer. By comparing the prediction results of FCN and full waveform inversion (FWI), the technical advantages are validated, and an in-depth analysis is conducted on the factors influencing the prediction accuracy of FCN. This effectively addresses the dual bottleneck of speed and accuracy encountered by traditional monitoring methods.

In response to the requirements for $CO_2$ migration monitoring based on physical property parameters, this chapter first establishes a solid theoretical foundation through numerical simulation of $CO_2$ migration, construction of a petrophysical model, and analysis of seismic response characteristics. An innovative fully convolutional neural network architecture based on multi-task learning is proposed, enabling simultaneous prediction of multiple formation parameters. The processes of training data preparation, network training, and synthetic data testing and validation are completed, ultimately achieving successful application to real data. This realises dynamic monitoring and quantitative characterisation of physical property parameters during the migration process, providing a reliable approach for in-depth analysis of migration patterns in sealed formations.

This chapter, through the innovation and practical application of two major technological systems, has overcome the efficiency and accuracy limitations in $CO_2$

geological storage migration monitoring. It provides core technical support for real-time dynamic management of storage projects, risk identification, and long-term safety assurance.

## References


1. Hornik K (1991) Approximation capabilities of multilayer feedforward networks. Neural Netw 4:251–257
2. Csaji B (2001) Approximation with artificial neural networks. Faculty of Sciences, Eötvös Loránd University
3. Burger HC, Schuler CJ, Harmeling S (2012) Image denoising: can plain neural networks compete with BM3D? In: IEEE computer society conference on computer vision and pattern recognition, pp 2392–2399
4. Jia B (2020) Prediction of protein and RNA binding sites based on fully convolutional neural networks. Jilin University, Changchun
5. Liu S (2020) Research on block masonry wall crack identification technology based on fully convolutional neural networks. Harbin Institute of Technology, Harbin
6. Yang F, Ma J (2019) Deep-learning inversion: a next generation seismic velocity-model building method. Geophysics 84(4):R583–R599
7. Dahl G, Sainath T, Hinton G (2013) Improving deep neural networks for LVCSR using rectified linear units and dropout. In: IEEE international conference on acoustics, speech and signal processing (ICASSP), pp 8609–8613
8. Ivandic M, Bergmann P, Kummerow J et al (2018) Monitoring $CO_2$ saturation using time-lapse amplitude versus offset analysis of 3D seismic data from the Ketzin $CO_2$ storage pilot site, Germany. Geophys Prospect 66(8):1568–1585
9. Gardner GHF, Gardner LW, Gregory AR (1974) Formation velocity and density; the diagnostic basics for stratigraphic traps. Geophysics 39(6):770–780
10. Köhn D, De ND, Kurzmann A et al (2012) On the influence of model parametrization in elastic full waveform tomography. Geophys J Int 191(1):325–345
11. Hassan W, Xi J (2012) Upscaling and its application in numerical simulation of long-term $CO_2$ storage. Greenh Gases Sci Technol 2(6):408–418
12. Mindlin RD (1949) Compliance of elastic bodies in contact. J Appl Mech 16:259–268
13. Hill R (1952) The elastic behaviour of crystalline aggregate. Proc Phys Soc Sect A 65(5):349–354
14. Ao Y, Lu W, Jiang B et al (2020) Seismic structural curvature volume extraction with convolutional neural networks. IEEE Trans Geosci Remote Sens 59(9):7370–7384
15. Li Y, Song J, Lu W et al (2021) Multitask learning for super-resolution of seismic velocity model. IEEE Trans Geosci Remote Sens 59(9):8022–8033
16. Wu X, Liang L, Shi Y et al (2019) Multi-task learning for local seismic image processing: fault detection, structure-oriented smoothing with edge-preserving, and seismic normal estimation by using a single CNN. Geophys J Int 210(3):2097–2109
17. Li D, Peng SP, Guo Y et al (2021) $CO_2$ storage monitoring based on time-lapse seismic data via deep learning. Int J Greenh Gas Control 108:103336
18. Akai T, Kuriyama T, Kato S et al (2021) Numerical modelling of long-term $CO_2$ storage mechanisms in saline aquifers using the Sleipner benchmark dataset. Int J Greenh Gas Control 110:103405
19. Li B, Li YE (2021) Neural network-based $CO_2$ interpretation from 4D Sleipner seismic images. J Geophys Res Solid Earth 126:e2021JB022524

# Chapter 7
# Integrated Well and Surface Microseismic Monitoring Technology for the Geological Storage Safety of Carbon Dioxide

At present, geophysical methods available for monitoring the safety of $CO_2$ geological storage include three-dimensional seismic surveys, microseismic monitoring, near-surface electromagnetic techniques, and satellite remote sensing, among others. This section primarily introduces the key technologies developed in recent years by the project team for the application of microseismic methods in monitoring the safety of $CO_2$ geological storage. The microseismic method monitors changes in formation pressure or microseismic events occurring during $CO_2$ injection, thereby reflecting the patterns of pressure variation as well as the migration of gas within the storage formation and aquifer. Acoustic signals generated during hydraulic fracturing or high-rate injection processes can be used to determine the locations of microseismic events.

Microseismic monitoring techniques primarily include two main approaches: downhole monitoring and surface monitoring. Downhole monitoring offers high vertical resolution, allowing for the avoidance or mitigation of the influence of low-velocity zones in the upper part of the section and environmental noise interference; however, it is relatively more challenging to implement. Surface monitoring, on the other hand, provides strong horizontal resolution and is comparatively easier to deploy, but compared to downhole monitoring, it typically yields fewer effective events, has a lower signal-to-noise ratio, and is more susceptible to external noise interference. Analysis indicates that downhole and surface monitoring methods are complementary; their combination can achieve better resolution in both horizontal and vertical directions, thereby improving the accuracy and capability of microseismic event localisation. Therefore, it is necessary to develop integrated downhole-surface microseismic monitoring technologies and equipment to enable high-precision monitoring in target areas.

The traditional microseismic processing workflow includes: (1) establishment of the velocity model; (2) calibration of the velocity model; (3) data pre-processing; (4) high-precision localisation; and (5) analysis of localisation results. Some steps



D. Li and Y. Guo, *Geological Storage and Monitoring of Carbon Dioxide*,
https://doi.org/10.1007/978-981-92-3621-3_7

in the workflow are relatively complex. For example, in data pre-processing, relevant tasks include weak signal extraction and microseismic event phase identification. Most traditional microseismic signal extraction methods lack adaptability and require manual threshold setting. For microseismic event phase identification under low signal-to-noise ratios, the process is mainly carried out manually with the support of specialised software. This not only results in low operational efficiency, but also means that the accuracy of event identification often depends on the operator's experience. In microseismic event localisation, particularly when using grid search methods such as the grid search algorithm, there are issues related to computational complexity.

In recent years, deep neural networks, with their highly complex representational capabilities, have found increasingly widespread application in the processing and interpretation of geophysical data. By introducing convolutional neural networks and generative adversarial networks, a deep network model framework for microseismic monitoring has been constructed. Research has been conducted into the relationship between fracturing and microseismic responses, and, by integrating deep learning theory, a bridge has been established between deep feature learning and microseismic response. Strategies such as transfer learning and semi-supervised learning have been employed to address the challenge of limited sample sizes in deep network training. Ultimately, this has enabled the development of an integrated intelligent processing workflow for microseismic records based on deep learning, encompassing noise suppression and signal enhancement, automatic event identification, accurate phase picking, and precise localisation of fracture events under specific monitoring systems. This provides technical support for fracture monitoring and addresses the issues of low efficiency in traditional manual data processing and the computational intensity of conventional processing methods.

## 7.1 Intelligent Preprocessing Methods for Microseismic Data

The preprocessing methods for microseismic data include two main steps: microseismic data filtering and initial picking of microseismic events. The signal strength of microseismic events is typically very weak, with a low signal-to-noise ratio, and is often submerged in noise. Prior to event location and seismic source mechanism inversion, it is necessary to preprocess the microseismic signals. The preprocessing methods described in this book cover two aspects: denoising and event picking/phase identification.

Common seismic data denoising methods often employ digital filters to directly process seismic data for noise reduction. By applying the Fourier transform, time-domain signals are projected into the frequency domain, where digital filters (such as FIR filters) can be designed as high-pass, low-pass, band-pass, or band-stop filters to attenuate noise components within the frequency domain. The denoised data are then obtained through inverse transformation. In addition to direct use of digital filters, other approaches such as f-k transform and wavelet transform can be utilised to decompose the signal, retaining the effective components of the signal while removing noise and other interfering components, thereby achieving denoising. Denoising methods based on empirical mode decomposition are also among the classic approaches; EMD/EEMD decomposes the signal into a series of intrinsic mode functions (IMFs) of different frequencies and a residual component (R), and denoising is achieved by processing the IMFs according to different objectives. The aforementioned traditional denoising methods are either overly reliant on manual intervention or are computationally intensive. Therefore, in response to the denoising problem, this study proposes two methods based on deep learning theory: one is a denoising method based on DnCNN employing a residual learning strategy; the other is a data-driven denoising method based on generative adversarial networks, aiming to achieve end-to-end microseismic data denoising.

The Short-Term Average/Long-Term Average (STA/LTA) energy ratio method is a classic algorithm for microseismic event detection, which identifies abrupt changes in seismic signals by comparing the ratio of the mean amplitude energy over short and long time windows. Subsequent research has demonstrated that replacing the mean amplitude in the time window with characteristic functions can effectively improve the accuracy of this method. However, this approach often performs poorly when dealing with low signal-to-noise ratio microseismic signals and is unable to distinguish between seismic phases (P-wave and S-wave) of events. To address this issue, Grigoli (2013) utilised the Hilbert transform based on three-component seismic data to enable the STA/LTA algorithm to extract both P-wave and S-wave phases. Tan Yuyang (2016) combined STA/LTA, polarisation analysis, and edge detection to achieve phase identification for microseismic events. The fundamental principle of energy ratio-based event detection lies in extracting signal features and identifying seismic events by tracking changes in these features. In addition to calculating the windowed energy, methods based on statistical models, such as the Akaike Information Criterion (AIC), as well as time–frequency domain transformation techniques, can also utilise waveform energy to detect microseismic events. However, as the demand for higher inversion accuracy increases, the detection and identification rates of traditional phase recognition methods are no longer sufficient to meet requirements. Therefore, in the detection and phase identification of microseismic events, this study proposes a lightweight network model based on SegUNet for three-component microseismic data by integrating convolutional neural networks and recurrent neural networks. For microseismic data signals that do not meet the three-component requirement, a phase identification algorithm based on SC-PSNET is proposed, in which the network input is single-component microseismic data.

### 7.1.1 Intelligent Filtering and Denoising Methods for Microseismic Data

When employing a sparse station deployment method for ground microseismic monitoring, the similarity between different channels is low. Therefore, using single-channel data for network training is more reasonable and can effectively reduce the risk of overfitting during training. Single-channel microseismic data can be regarded as the superposition of effective signals and noise.

$$x = s + n \tag{7.1}$$

Here, $x$ represents the actual microseismic channel data, $s$ denotes the signal to be extracted, and $n$ refers to the noise signals to be filtered out, including both random noise and coherent noise signals. The feedforward process of a deep convolutional neural network can be abstracted as follows: $y_{out}$ is the output of the network, $x_{in}$ is the input to the network, $y_{out}$ represents the parameters of the network, and $\theta$ denotes the nonlinear generative network that maps $f$ to $x_{in}$.

$$y_{out} = f_\theta(x_{in}) \tag{7.2}$$

According to the different training methods, the use of neural networks for effective signal reconstruction can be categorised into supervised and unsupervised approaches. Based on different learning strategies, these can be further divided into waveform learning and residual noise learning.

Using a supervised learning approach, different learning strategies can be abstracted for the network as the following formula:

$$\hat{x} = f_{\theta 1}(y) \quad \theta_1 = arg \min_\theta loss(\hat{x}, x) \tag{7.3}$$

$$\hat{n} = f_{\theta 2}(y) \quad \theta_2 = arg \min_\theta loss(\hat{n}, y - x) \tag{7.4}$$

$\hat{x}$ represents the distribution of effective signals predicted by the network, while $\hat{n}$ denotes the distribution of noisy signals predicted by the network. $\theta_1$ and $\theta_2$ are the optimal network parameters found by minimizing the loss function. The loss function is typically constructed using the root mean square error, as shown in the formula below:

$$loss(y_{out}, y_{label}) = f_\theta(x_{in}) - y_{label}^2 \tag{7.5}$$

Among them, $y_{label}$ represents the label data, which needs to be adjusted accordingly based on the specific learning strategy employed.

(1) Microseismic Data Denoising Method Based on Generative Adversarial Networks

By introducing a generative adversarial network, this section proposes an end-to-end denoising method for single-channel microseismic data. The network model comprises two components: a generator and a discriminator, whose structures are shown in Figs. 7.1 and 7.2, respectively. The generator is constructed to produce processed data, and after the network training is completed, it reconstructs the denoised data. The discriminator is responsible for distinguishing between noisy and noise-free data. The generator network G adopts a fully convolutional encoder-decoder architecture. Compared with traditional fully convolutional encoder-decoder structures, it does not employ max pooling for downsampling, but instead uses strided convolution operations with a stride greater than one, which provides enhanced stability. The discriminator D consists of a convolutional encoder part and a fully connected layer part.

The generator G and discriminator D are trained through adversarial propagation. Unlike traditional deep convolutional neural networks, the training of the GAN network is carried out in three steps:

(i) Use both noisy and noise-free data as input to train the discriminator, and label the output as "true";
(ii) Use the data generated by the generator and the clean data as input, label the output as "fake", and retrain the discriminator;
(iii) Freeze the discriminator parameters, use the generated data and clean data as input, and label the output as "true" to train the generator.

With alternating iterative training, both the generator and discriminator networks become increasingly powerful. After training, the generator is able to reconstruct

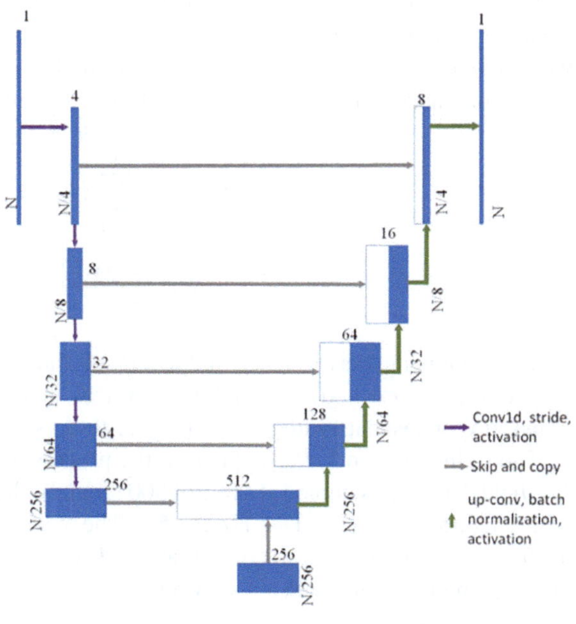


**Fig. 7.1** Architecture of the generator network G

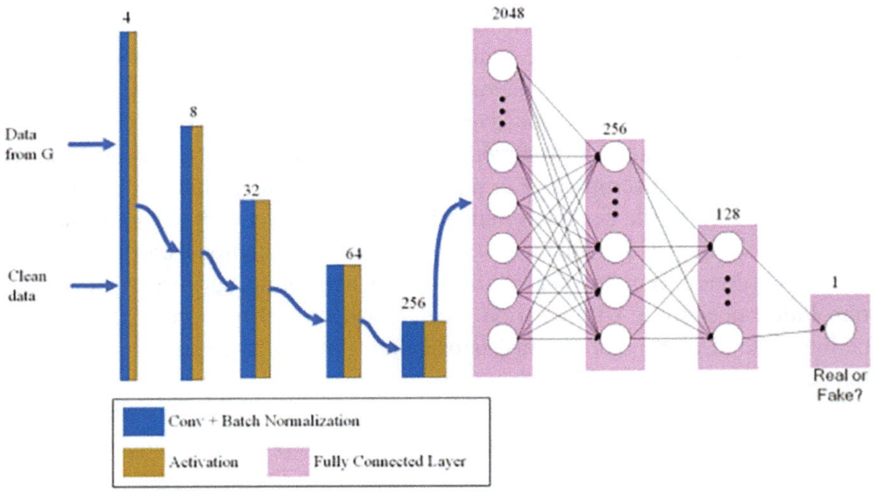


**Fig. 7.2** Architecture of the discriminator network D

the actual signal distribution and produce denoised data similar to clean data. The discriminator will be unable to distinguish between real clean data and the denoised data generated by the generator.

Microseismic signals without noise were obtained through forward modelling, and subsequently divided into training and testing sets. Both the training and testing data are synthetic, but are generated using different numerical models. The original synthetic signals are free from noise interference. For the purposes of network training and testing, Gaussian white noise of varying intensities was added to the original synthetic signals. To quantitatively assess the denoising performance of the method, we employed the signal-to-noise ratio (SNR) and root mean square error (RMSE) as evaluation metrics. The specific forms of these two metrics are as follows:

$$MSE = \frac{1}{N}\sum_{n=1}^{N}(s_{Error})^2 \tag{7.6}$$

$$R = \frac{cov(s, \hat{s})}{\sigma_s \times \sigma_{\hat{s}}} \tag{7.7}$$

The denoising performance indicators of the GAN-based denoising method and traditional methods (ensemble empirical mode decomposition and wavelet transform) on test data with different signal-to-noise ratios are compared in Fig. 7.3.

The trained network was applied to the processing of actual microseismic data. The field data were sampled at a rate of 500 Hz and collected by 12 surface geophones with a natural frequency of 4.5 Hz ± 7.5%. The observation system was designed and deployed in a sparse station configuration. The results of the network's processing of actual microseismic data are shown in Fig. 7.4, demonstrating good performance.

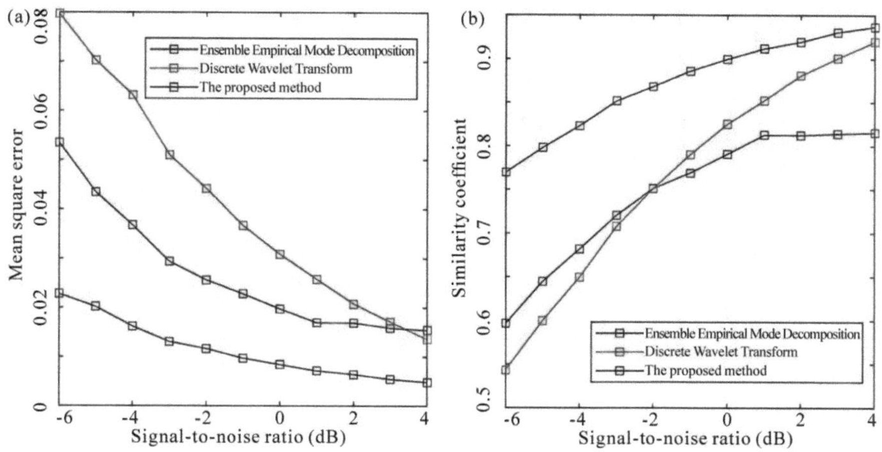


**Fig. 7.3** Analysis of denoising performance under different SNRs: **a** MSE analysis; **b** cross-correlation coefficient analysis

Compared with traditional microseismic signal extraction methods, the proposed approach exhibits superior results.

(2) Microseismic Data Denoising Method Based on DnCNN

This section proposes a denoising method based on DnCNN, which was initially applied to the processing of random noise in images. Unlike other CNN-based denoising approaches, this method employs a residual learning strategy, whereby the neural network predicts the noise signal rather than the denoised signal. Practical validation demonstrates that adopting the residual learning strategy results in more stable network training and improved denoising performance. After network training is completed, the denoised data can be obtained by subtracting the noise predicted by the network from the original measured data. The method was tested using both simulated and real data, and experimental results indicate that the approach proposed in this section achieves higher efficiency and better results compared to commonly used denoising methods (Fig. 7.5).

(i) Methodology and Process Flow Diagram

The basic framework of the neural network proposed in this section is shown in Fig. 7.6. As illustrated, the input to the neural network is a seismic signal with noise, and the output of the network is the extracted noise signal.

(ii) Experimental Data and Test Results

The following section utilises actual data to test the denoising effectiveness of the proposed methods. The real data were collected by ground-based detectors with a common resonance frequency of 4.5 Hz $\pm$ 75%, and the detectors were not positioned along the same straight line on the ground. The data sampling interval

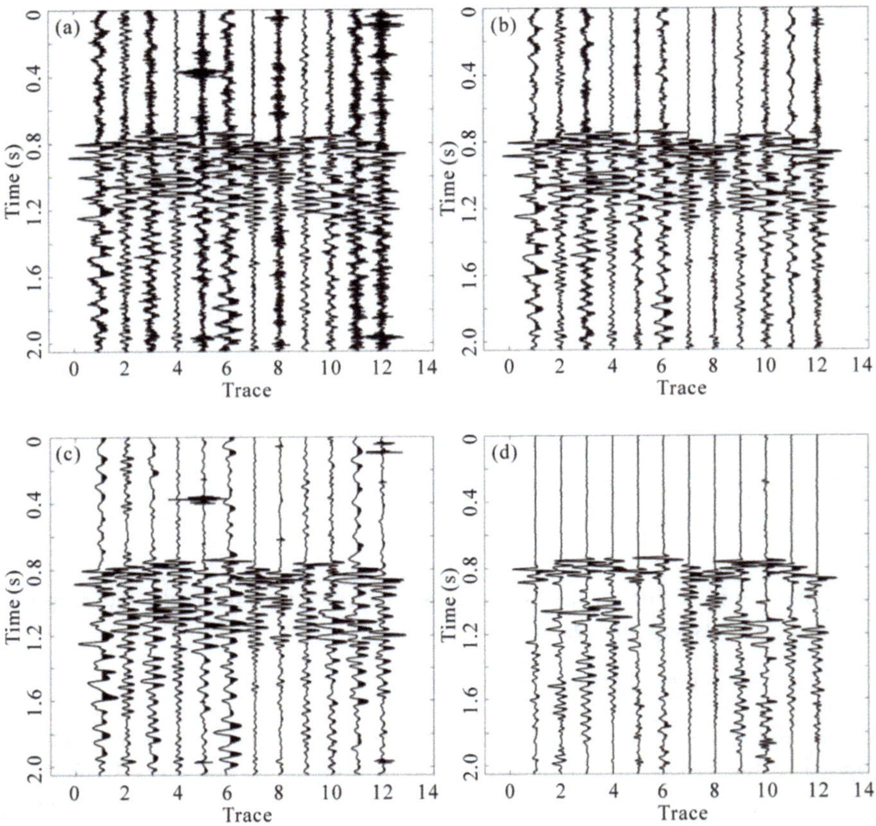


**Fig. 7.4** Comparison of denoising effects of different methods on the field dataset: **a** field data; **b** denoising result by ensemble empirical mode decomposition; **c** denoising result by wavelet transform; **d** denoising result by GAN

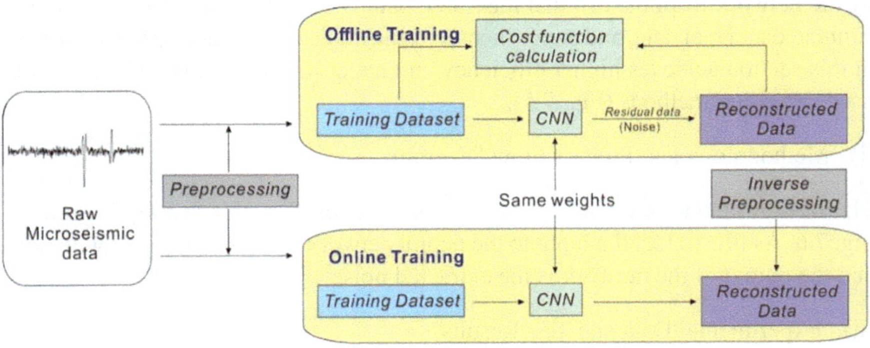


**Fig. 7.5** Workflow of the denoising method

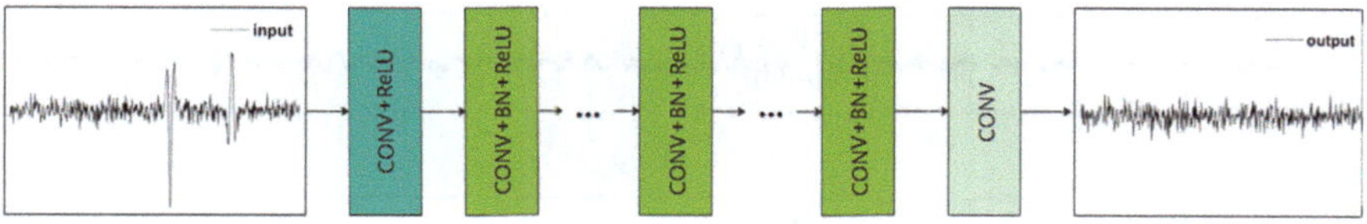


**Fig. 7.6** Neural network architecture of the proposed method

was. Two sets of real data were used to evaluate the denoising performance: the first dataset contains only a single microseismic event, while the second dataset includes multiple microseismic events. Three methods—namely, a neural network-based approach, an ensemble empirical mode decomposition-based approach, and a wavelet decomposition-based approach—were employed to denoise the real signals. The denoising results for the three methods applied to the real data containing only one microseismic event are shown in Fig. 7.7. It can be observed that the method proposed in this section almost completely removes all noise signals, resulting in a very "clean" signal prior to the microseismic event. This is highly beneficial for the extraction and phase identification of microseismic events.

Experimental results indicate that, whether applied to synthesised data or to microseismic data collected in the field, the method proposed in this section can effectively suppress noise present in the original signals. It is capable of separating almost all noise signals from the original data, resulting in denoised data that contain "clean" microseismic event waveforms that are virtually unaffected by noise interference. This provides a solid foundation for subsequent microseismic event detection and phase identification.

### *7.1.2 Intelligent First Break Picking for Microseismic Data*

(1) Seismic Phase Identification Method Based on SegUNet

In surface microseismic monitoring tasks, due to the sparse spatial distribution of sensors within the monitoring area, the recordings are not spatially continuous but are instead continuous in time. Therefore, a single-station microseismic phase identification model is designed based on a convolutional neural network, using three-component data as input. The network output also consists of three channels, representing the probability distributions of P-wave onset, S-wave onset, and noise, respectively. The training data for the network are obtained through limited-difference forward modelling in the time domain.

SegUNet is designed as a combination of SegNet and UNet. Considering the importance of spatial positions in feature layers for segmentation tasks, SegUNet introduces the "index pooling" operation from SegNet, which records the index position information of features during the pooling process. This approach avoids the loss of boundary feature information and, at the same time, utilises the index information during feature upsampling to restore boundary features. The operational details are

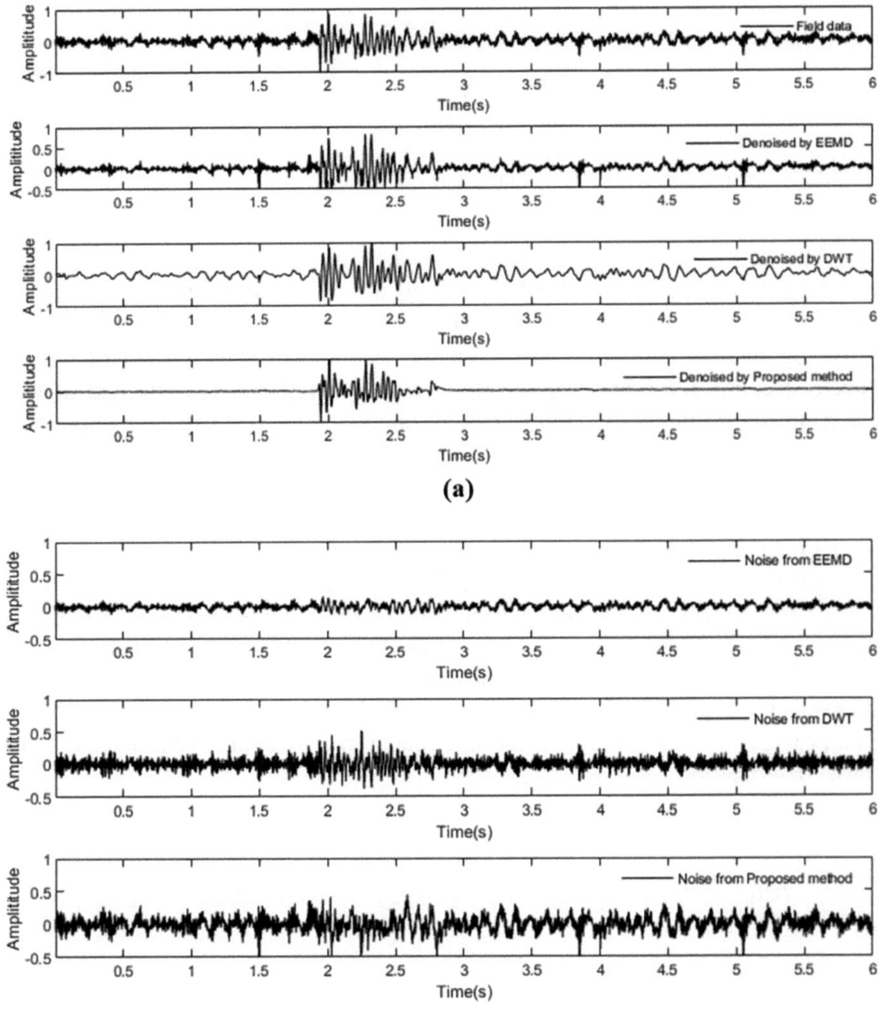


(a)

(b)

**Fig. 7.7** Denoising results of three methods for the first trace (containing only one microseismic event): **a** field data and denoising results; **b** noise signals extracted from the denoising results

shown in Fig. 7.8. Meanwhile, SegUNet retains the "skip and copy" operation from UNet, which is used for the fusion of high- and low-level features. In summary, the combined strategy of SegUNet with respect to UNet and SegNet is mainly reflected in the fusion of high- and low-level features, as well as the preservation of boundary features in the pooling and upsampling stages.

In terms of overall parameters, the encoder network is a typical convolutional network with a depth of four layers. Unlike the dual-convolutional structure of the UNet network, at each encoding depth, only a single convolutional block (including

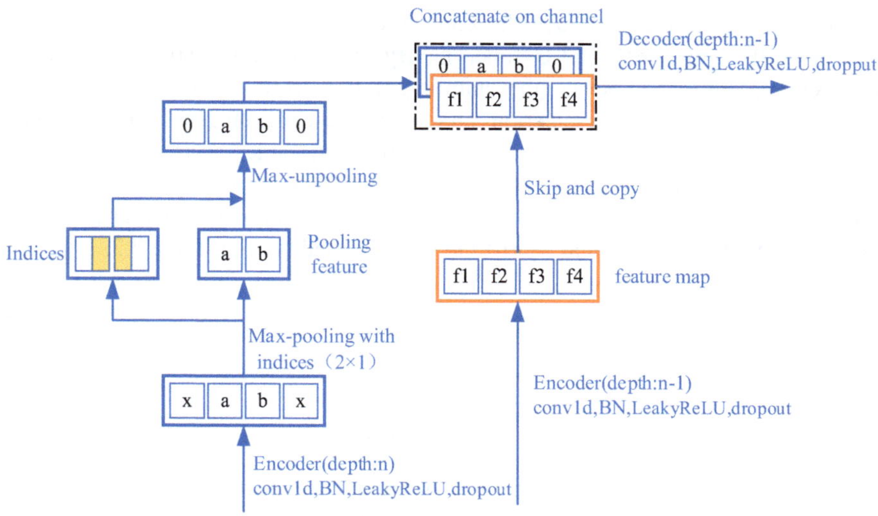


**Fig. 7.8** Illustration of the combination strategy of SegNet and UNet

convolution and pooling) is present. Additionally, the upsampling section adopts a transposed convolution structure, further reducing the number of parameters. The total number of parameters in the network is approximately 13,000, which enables faster inference efficiency. The overall network architecture is shown in Fig. 7.9.

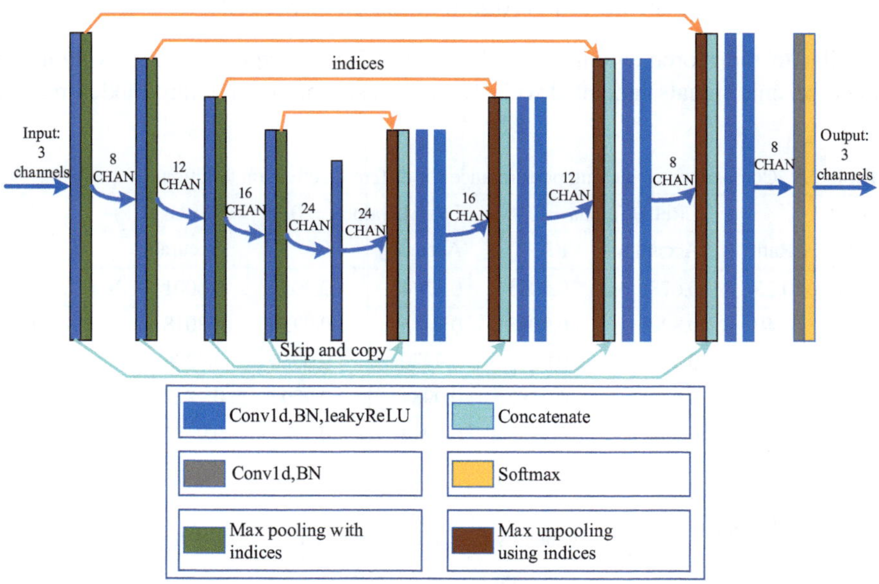


**Fig. 7.9** Design of the SegUNet network architecture

SegUNet no longer employs the conventional ReLU activation function, but instead utilises the LeakyReLU function. By retaining information in the negative region, this can enhance the stability of the given model. The impact of different activation function selections on seismic phase picking results under varying signal-to-noise ratios in synthetic datasets is shown in Table 7.1.

On the test set, a detailed analysis of the extraction performance of SegUNet was conducted using the precision P and recall rate R, and a comparative analysis was performed with the traditional short-time window method. See Table 7.2.

$$\mathrm{P} = \frac{TP}{TP + FP} \tag{7.8}$$

$$\mathrm{R} = \frac{TP}{TP + FN} \tag{7.9}$$

Among them, *TP* denotes the number of correctly identified seismic phases, *FP* represents the number of incorrectly identified seismic phases, and *FN* indicates the number of missed seismic phases.

The comparative results of actual microseismic event processing are shown in Fig. 7.10. The red solid line represents the probability distribution of the P-wave first arrival, while the green solid line represents the probability distribution of the S-wave first arrival. The lower part of each subfigure shows the picking results of P-STA/LTA and S-STA/LTA, and the upper part displays the picking results of SegUNet. Figure 7.10a–l correspond to the picking results for different microseismic events.

(2) Seismic Phase Identification Algorithm Based on SC-PSNET

Similar to the aforementioned SegUNet, SC-PSNET acquires a large volume of microseismic signals annotated with P and S phases through forward modelling and

**Table 7.1** Comparison of picking performance for different activation functions

| SNR | | LReLU (alpha = 0.3) | | LReLU (alpha = 0.1) | | ReLU | |
|---|---|---|---|---|---|---|---|
| dB | Ratio | Accuracy | F1 | Accuracy | F1 | Accuracy | F1 |
| − 6 | 0.25 | 0.6785 | 0.8085 | 0.8274 | 0.9055 | 0.6001 | 0.7501 |
| − 4 | 0.40 | 0.8296 | 0.9068 | 0.9475 | 0.9730 | 0.8018 | 0.8900 |
| − 2 | 0.63 | 0.9071 | 0.9513 | 0.9920 | 0.9960 | 0.9226 | 0.9597 |
| 0 | 1 | 0.9383 | 0.9682 | 0.9984 | 0.9992 | 0.9827 | 0.9913 |
| 2 | 1.58 | 0.9581 | 0.9786 | 0.9993 | 0.9997 | 0.9955 | 0.9978 |

**Table 7.2** Comparative analysis of picking results on the test set

| Contents | SegUNet | STA/LTA |
|---|---|---|
| Precision | 0.9448 | 0.4392 |
| Recall | 0.7409 | 0.4519 |

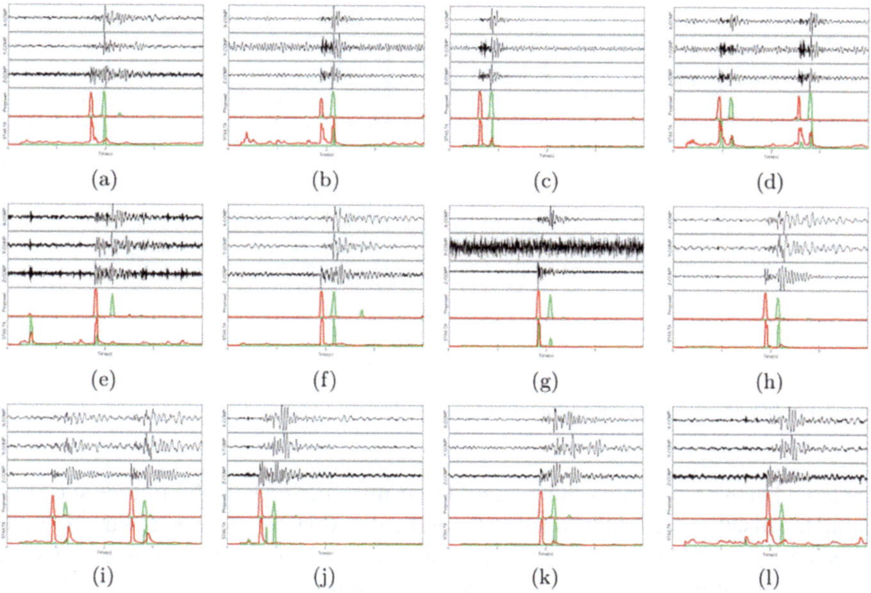


**Fig. 7.10** Comparison of picking results for actual microseismic data

data augmentation. However, unlike SegUNet, SC-PSNET introduces different vp/vs scenarios during the data simulation process to enhance the phase picking accuracy of the CNN-based model. Moreover, SC-PSNET is capable of performing phase picking using only single-component microseismic data, thereby offering broader applicability. At the same time, by integrating recurrent convolutional neural networks, it overcomes the challenge of insufficient feature representation in single-component records.

The network structure of SC-PSNET is shown in Fig. 7.11. The figure presents the processing results of single-channel data after the completion of network training, alongside a comparative analysis with STA/LTA and PhaseNet (Fig. 7.12).

## 7.2 Intelligent Localisation Methods for Microseismic Data

Microseismic source localisation methods can be categorised according to their underlying principles into: travel-time-based localisation methods, amplitude stacking-based localisation methods, and reverse-time migration-based localisation methods. Travel-time-based algorithms for microseismic event localisation often suffer from reduced accuracy due to the difficulty in obtaining precise travel times from low signal-to-noise ratio signals and inaccuracies in the velocity model. Furthermore, travel-time-based methods neglect amplitude and polarity information

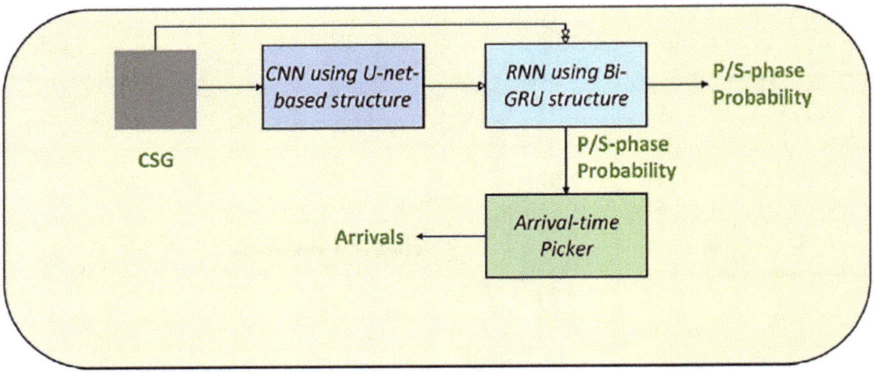


**Fig. 7.11** SC-PSNET network architecture

contained in microseismic signals, resulting in information loss. Amplitude stacking-based localisation methods utilise the theoretical travel times between grid points and detectors to stack amplitudes; the location corresponding to the maximum stacked energy is identified as the source location. Amplitude stacking-based localisation methods do not impose strict requirements on the picking accuracy of microseismic events and exhibit a certain degree of robustness to low signal-to-noise ratios.

### 7.2.1 *Automatic Microseismic Location Method Based on Phase Identification and Amplitude Stacking*

However, the waveform stacking-based localisation method still needs to address the following issues: (1) As the origin time is unknown, the localisation method requires searching in a four-dimensional space (time and hypocentre location), which is computationally intensive, or alternatively, manual/semi-automatic extraction of events and phase identification from seismic data is needed; (2) automatic localisation algorithms require phase identification; (3) the radiation pattern generated by the source mechanism may result in "zero stacking" of waveforms at the source location.

To address the aforementioned issues, this book proposes a localisation method based on deep learning for microseismic phase identification combined with waveform stacking. In order to avoid searching for the origin time during the stacking of amplitudes, the method first selects the data from a particular detector as the "reference channel", and then performs phase identification on the events in this reference channel. The picking of events in the remaining seismic channels is accomplished by calculating the theoretical travel time differences between channels based on grid points. To prevent "zero stacking" of amplitude energy at the source location due to random noise, the method exploits the distinct polarity of waveform cross-correlation prior to waveform stacking.

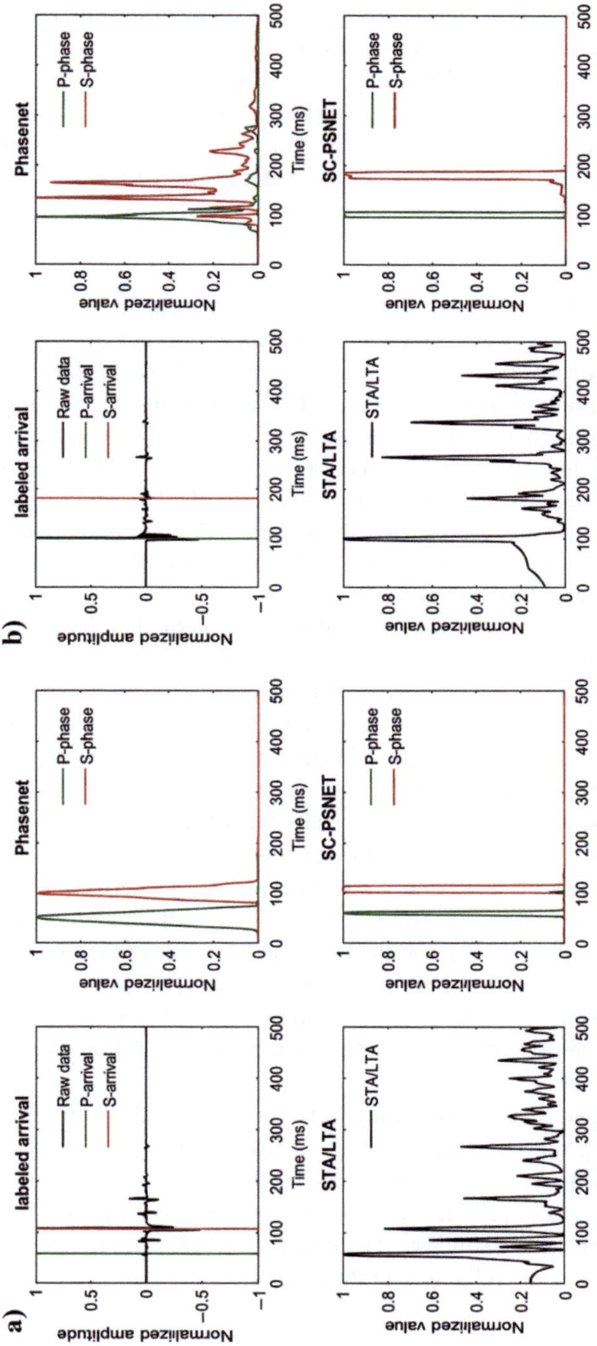


**Fig. 7.12** Picking analysis of actual data

(1) Principle and Procedure of the Method

A localisation method combining microseismic phase identification based on deep learning with waveform stacking is implemented in three steps: (a) the monitoring area is divided into a grid and the theoretical travel times of the detectors corresponding to each grid point are calculated; (b) the seismic phases of the reference channel data for seismic events are identified; (c) automatic localisation imaging is achieved by stacking according to the theoretical travel times and the relative amplitudes of the identified phases.

(2) Actual Test Results

To evaluate the capability of the localisation algorithm in processing real-world data, actual microseismic data from a certain region were selected, comprising a total of 34 sets of three-component data, sampled at a frequency of 5000 Hz, with each dataset containing 6000 sampling points (Fig. 7.13). The 34 detectors were arranged on the surface, with their distribution shown in Fig. 7.14. The P-wave velocity is displayed, while the S-wave velocity is calculated according to the ratio *TP*. The horizontal grid points are divided as shown in the figure, with a spacing of 20 m between discrete grid points. Vertically, the grid is discretised from 2700 to 3400 m at intervals of 20 m. Upon completion of the grid division, the travel-time lookup table is calculated based on the given P-wave and S-wave velocities.

After the lookup table calculation is completed, the method proposed in this book is employed to automatically locate microseismic events. Firstly, a deep learning-based microseismic phase identification method is used to extract and identify P-wave and S-wave events from the reference channel data. The "tilted channel" in the original data is removed before inputting the data into the trained neural network. The results of microseismic phase identification are shown in Fig. 7.15. As can be seen from the figure, the phase identification results for the actual data are satisfactory. The first seismic channel is selected as the reference channel, and the S-wave event is chosen for localisation using the y-component data. To verify the localisation effectiveness, commercial microseismic data processing software is used to locate the event under the same conditions, and the localisation results of the two methods are shown in Fig. 7.16. Although the grid partitioning differs between the two methods, the localisation results at the same depth in the x and y planes are similar, further confirming the reliability and accuracy of the localisation algorithm proposed in this section.

In summary, this study demonstrates that the proposed localisation method achieves accurate positioning and exhibits a certain degree of robustness against noise, velocity errors, and event timing errors. Furthermore, the method utilises real-world data for localisation, and the results have been compared with those obtained from established processing software. The localisation outcomes of both approaches are similar, confirming the reliability of the proposed algorithm.

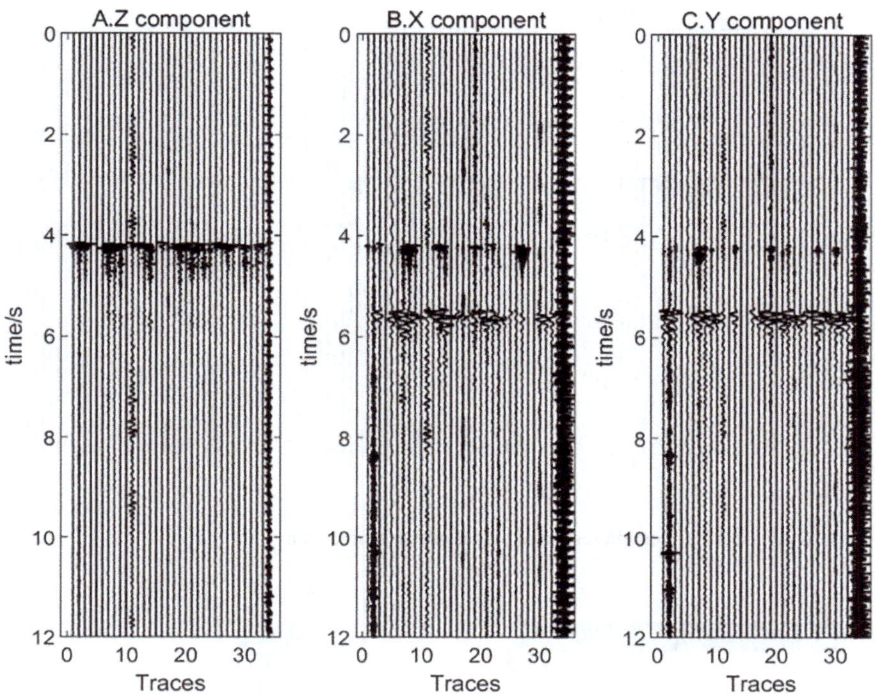


Fig. 7.13 Actual microseismic data (three components)

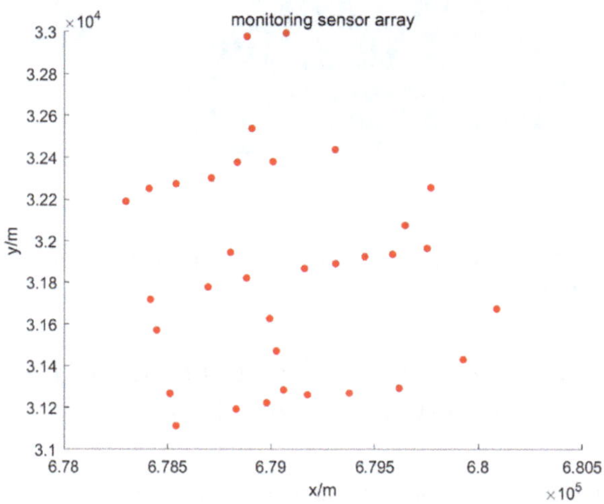


Fig. 7.14 Distribution of surface geophones

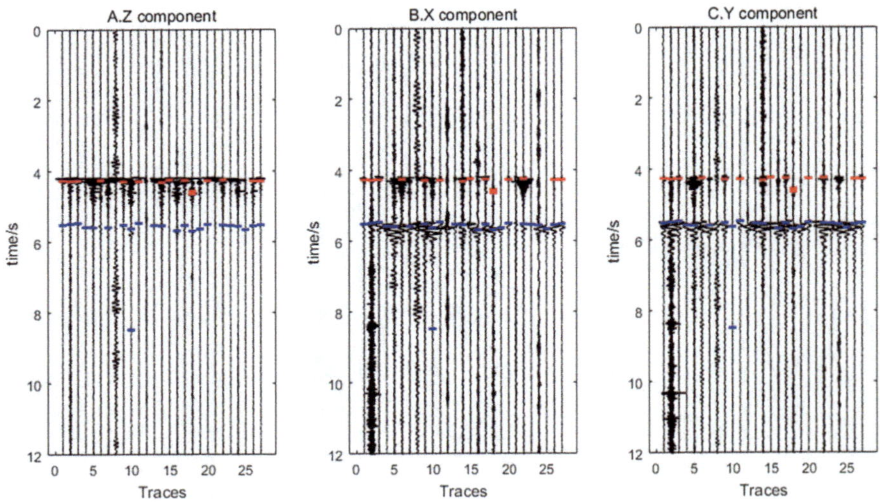


**Fig. 7.15** Phase-identification results of actual microseismic data

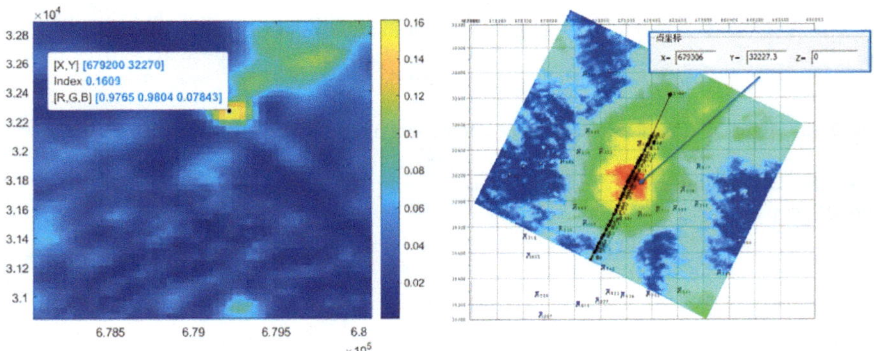


**Fig. 7.16** Comparison of location results: **a** results of the proposed method; **b** results of commercial software

## *7.2.2 Method for Microseismic Event Location Based on ResNet Network Model*

The automatic localisation method based on waveform stacking can utilise the amplitude of microseismic data for localisation. The method discussed above employs only a single seismic component for waveform localisation after phase identification. Although it is possible to stack data from all three components, this would result in a twofold increase in the number of localisation events, thereby affecting localisation efficiency.

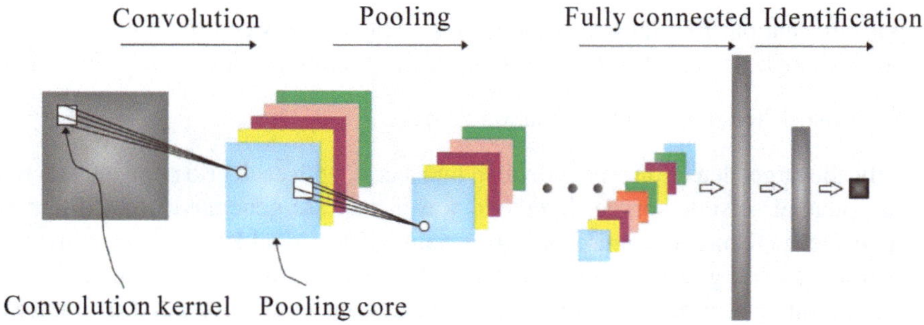


**Fig. 7.17** CNN network architecture

In recent years, the field of deep learning has demonstrated remarkable capabilities in computer vision, with existing networks achieving classification accuracies of over 90% in multi-class image classification tasks. Since the AlexNet network achieved victory in image classification competitions, deep residual networks have attracted significant attention. Subsequently, ResNet has become prominent for addressing the degradation problem in deep neural networks, enabling the training of networks with thousands of layers and delivering robust performance.

(1) Principle of the Method

ResNet is a type of deep neural network based on CNN (Convolutional Neural Network). In image processing, if image data are vectorised and then processed by a fully connected network, it can result in an excessive number of neural network parameters, making training unfeasible. CNNs reduce the number of parameters to be trained by employing shared convolutional kernels, thereby enhancing the generalisation capability of the model. A typical CNN network structure, as illustrated in Fig. 7.17, consists of an input layer, convolutional layers, activation layers, pooling layers, fully connected layers, and an output layer.

The input layer of a CNN consists of either two-dimensional or three-dimensional neurons. For grayscale images, the input comprises two-dimensional neurons, while for colour images (RGB images), it consists of three-dimensional neurons.

The convolutional layer is composed of convolutional kernels (filters). The convolutional layer generates feature maps by convolving the input data with these kernels. The size and number of convolutional kernels are typically determined according to the dimensions of the input data. The size of the convolutional kernels determines the dimensions of the feature maps, while the number of kernels determines the number of feature maps produced.

To handle increasingly complex tasks, the number of layers in neural networks has gradually increased. However, experimental results have shown that the performance of neural networks does not always improve accordingly; instead, a downward trend may occur. The ResNet network effectively addresses the problem of network degradation. Unlike conventional CNN-based networks that directly learn the features of input information, ResNet learns the residual features between the output and the input. Its core lies in the introduction of the "identity shortcut connection" (Fig. 7.18).

(2) Network Training and Localisation Process

Firstly, the target detection area is divided into a grid, with each grid point considered as a potential seismic source. Synthetic waveforms are generated using different source mechanisms, and noise of varying intensities is added to these synthetic waveforms. Each grid point index is used as a label for network training. Due to the different source mechanisms, a single grid point may correspond to multiple sets of seismic data. As most microseismic data consist of background noise and microseismic events are sparse within the event domain, it is necessary to perform down-sampling and truncation on the synthetic seismic data. To ensure that the

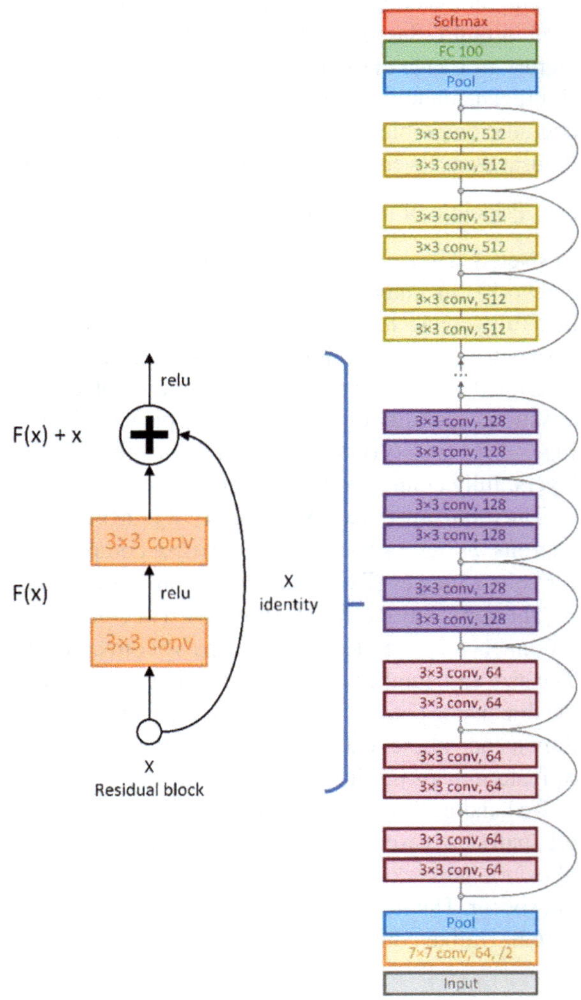


**Fig. 7.18** Basic structure of ResNet

truncated synthetic seismic data retain all amplitude information of the events as well as the amplitude curve characteristics, the phase identification method proposed in this study can be used after down-sampling to align the P-wave peaks at the same position. In this way, the data will contain all event waveform information while preserving the relative relationships between amplitude curves.

The normalised data with added noise are ultimately divided into two categories: a training set and a test set. The training set is used to train the neural network. After training, the network classifies the three-dimensional data from the test set, and the classification results can be mapped to a specific location outcome, thereby achieving the objective of localisation. The detailed localisation process is shown in Fig. 7.19.

(3) Analysis of Experimental Results

To verify the effectiveness of the localisation method, a numerical model was established to generate synthetic data, as illustrated in Fig. 7.20. The detector array is arranged on the surface (red dots in the figure), with the range in the $x$ direction being $(-1.5, 1.5)$ km, and in the $y$ direction being $(0, 3)$ km, comprising a total of 224 detectors. A target detection area is designed at a depth of 2.2 km, randomly

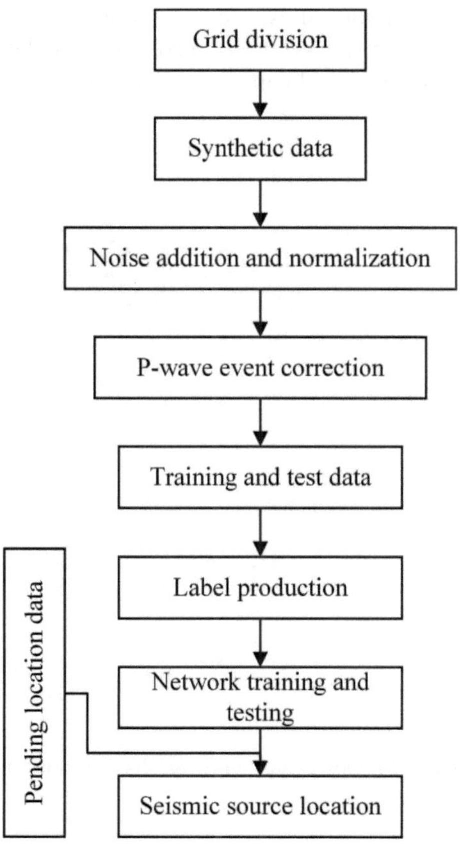


**Fig. 7.19** Workflow of the location algorithm

divided into 100 grid points (blue dots in the figure). The grid points can be divided either uniformly or randomly. The velocity model, as shown in Table 7.3, is a four-layer horizontal stratified model, with 512 sampling points, a sampling interval of 0.004 s, and a dominant frequency of 10 Hz. At each grid point, synthetic records are calculated using the forward modelling programme with different source mechanisms. The source mechanism is characterised by three parameters of a double-couple shear dislocation source (strike, dip, and rake), which are discretised into three parameter sets for simplification. When generating synthetic data, the strike parameter is $\{25°, 30°, 35°, 40°, 45°\}$, the dip parameter is $\{55°, 60°, 65°, 70°\}$, and the rake parameter is $\{160°, 165°, 170°, 175°, 180°\}$. The combination of these three parameters yields 100 types of source mechanisms, corresponding to 100 radiation patterns. With 100 random grid points, each grid point corresponds to 100 radiation patterns, resulting in a total of 10,000 sets of seismic data. Noise levels of – 5, – 2.5, and 0 dB are added to the data, producing a total of 40,000 sets of synthetic seismic data. After the synthetic data are generated, a down-sampling process is applied to obtain 224 sampling points, resulting in three-component seismic data with a size of $(3 \times 224 \times 224)$. Figure 7.21 shows the synthetic waveforms obtained at the same grid point using different source mechanisms. By comparison, it can be seen that the waveform curves of the seismic records are identical, but the amplitude energy distribution varies, and the polarity of the amplitude extrema may be reversed.

After the synthesis record is completed, 40,000 earthquake data entries are assigned a grid partition serial number based on the coordinates of the grid points.

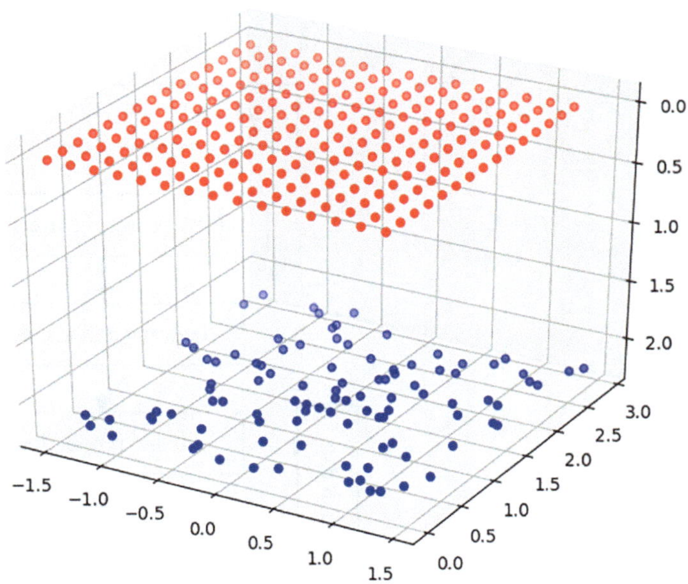


**Fig. 7.20** Observation array (red dots) and randomly partitioned grids (blue dots)

**Table 7.3** Layered velocity model

| Layer number | Depth (km) | P-wave velocity (km/s) | P/S velocity ratio |
|---|---|---|---|
| 1 | 0.08 | 0.90 | 1.730 |
| 2 | 0.32 | 1.60 | 1.731 |
| 3 | 0.95 | 2.125 | 1.731 |
| 4 | 1.65 | 2.30 | 1.730 |

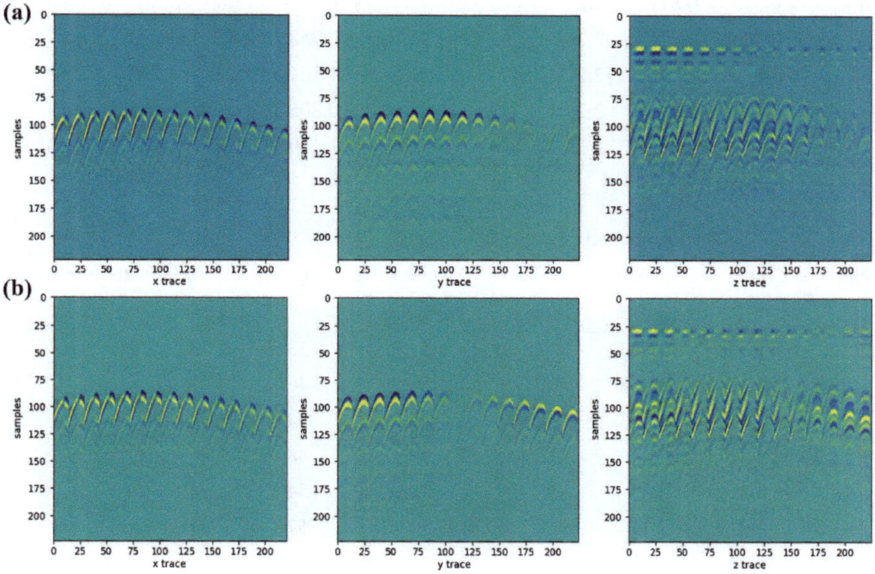


**Fig. 7.21** Synthetic records generated at the same source location by different source mechanisms

This serial number is then used as the label for each earthquake data entry. The correspondence of some labels is shown in Table 7.4.

To date, 40,000 synthetic earthquake records have been generated and labelled, and their order has been randomised. The training and testing datasets are allocated in a 7:3 ratio. For the network training parameters, the batch size is set to 20 and the number of epochs is 30.

After the completion of network training, 12,000 test set samples were input into the network to evaluate its classification accuracy. The results showed that the network achieved a classification accuracy of 100%, indicating that the well-trained neural network correctly classified each seismic data sample to its corresponding true seismic source location. Some of the classification results are shown in Table 7.5.

Experimental results show that for 12,000 sets of seismic data, the well-trained network accurately maps all of them to their corresponding true spatial locations, achieving a classification accuracy of 100%. The neural network is capable of successfully handling the classification of 100 spatial locations, with each location

**Table 7.4** Selected grid-point positions and corresponding labels

| No. | x (km) | y (km) | z (km) | Label | No. | x (km) | y (km) | z (km) | Label |
|---|---|---|---|---|---|---|---|---|---|
| 1 | – 1.041 | 2.149 | 2.2 | 1 | 11 | 0.597 | 0.965 | 2.2 | 11 |
| 2 | 1.491 | 2.68 | 2.2 | 2 | 12 | 0.267 | 2.607 | 2.2 | 12 |
| 3 | 1.326 | 0.248 | 2.2 | 3 | 13 | 1.18 | 2.965 | 2.2 | 13 |
| 4 | – 0.514 | 0.045 | 2.2 | 4 | 14 | – 0.771 | 0.328 | 2.2 | 14 |
| 5 | 0.377 | 1.818 | 2.2 | 5 | 15 | 0.138 | 1.242 | 2.2 | 15 |
| 6 | – 0.315 | 0.817 | 2.2 | 6 | 16 | – 0.82 | 1.689 | 2.2 | 16 |
| 7 | 0.105 | 2.15 | 2.2 | 7 | 17 | – 0.958 | 2.428 | 2.2 | 17 |
| 8 | 1.337 | 1.975 | 2.2 | 8 | 18 | 1.101 | 1.855 | 2.2 | 18 |
| 9 | – 0.596 | 2.57 | 2.2 | 9 | 19 | – 0.805 | 1.983 | 2.2 | 19 |
| 10 | 1.178 | 2.337 | 2.2 | 10 | 20 | 1.149 | 2.861 | 2.2 | 20 |

**Table 7.5** Neural-network-predicted location results

| No. | x (km) | y (km) | z (km) | Noise (dB) | Label | Classification result |
|---|---|---|---|---|---|---|
| 1 | 1.337 | 1.975 | 2.2 | – 5 | 7 | 7 |
| 2 | 1.359 | 1.543 | 2.2 | – 5 | 33 | 33 |
| 3 | – 0.376 | 2.228 | 2.2 | – 5 | 40 | 40 |
| 4 | – 1.308 | 1.183 | 2.2 | – 5 | 44 | 44 |
| 5 | 0.971 | 0.436 | 2.2 | – 5 | 48 | 48 |
| 6 | – 0.85 | 2.23 | 2.2 | – 5 | 54 | 54 |
| 7 | 0.011 | 2.274 | 2.2 | – 5 | 57 | 57 |
| 8 | – 0.838 | 1.39 | 2.2 | – 5 | 96 | 96 |

corresponding to 100 samples exhibiting different radiation patterns, and the data are further subjected to varying degrees of noise interference. The results presented in the table are based on data with – 5 dB noise interference, and all classification outcomes are correct, indicating that the method proposed in this section achieves accurate localisation and demonstrates strong robustness to noise interference. In the localisation process, this method utilises the complete set of three-component data from all microseismic events, without the need to consider the impact of event magnitude on localisation results. The trained neural network can directly map seismic data, even when affected by different levels of noise interference, to the corresponding classification labels, with each classification label corresponding to a spatial location, which constitutes the localisation result.

After training the neural network with a dataset of 12,000 samples, the network was tested. The results indicate that the network correctly mapped all seismic data in the 12,000-sample test set to their respective epicentre locations, achieving a localisation accuracy of 100%. The advantages of this method are as follows: (1) it utilises three-component data for localisation, while simultaneously making use of information from both P-waves and S-waves; (2) the training process of the network takes into account the influence of radiation patterns, thus there is no need to consider the effects of amplitude polarity or amplitude strength on localisation during application; (3) the training dataset includes the influence of noise with varying intensities, so the localisation algorithm demonstrates a certain degree of robustness to noise interference while ensuring the correctness of localisation results.

### 7.2.3 *Deep Learning-Based Joint Inversion Method for Microseismic Event Location and Velocity Model*

To achieve the goal of joint inversion of velocity models and source locations, two main network models were constructed. Network I utilises seismic waveform data to predict the subsurface velocity structure model. Network II combines the original seismic waveform records with the velocity model information inverted by Network I as inputs to the network, thereby performing inversion and localisation of the earthquake source. The accuracy of the localisation results is ensured by incorporating constraint information on the velocity of the mining area. This method can be applied to mining areas with different subsurface velocity structures, demonstrating strong generalisation capability and broad applicability. The network structure is shown in Fig. 7.22.

The network architecture employed is the Fully Convolutional Network (FCN). The FCN was first proposed by Long Jonathan and colleagues as a method for semantic segmentation. In contrast to Convolutional Neural Networks (CNNs), the FCN replaces the fully connected layers in traditional CNNs with convolutional layers. The entire network is thus composed exclusively of convolutional layers, hence the term "fully convolutional network." A basic FCN architecture consists of convolutional layers, pooling layers, and deconvolutional layers. The fully convolutional network can be represented as follows:

$$y = Net(x; \vartheta) = S(K_2 * (M(R(K_1 * x + b_1))) + b_2) \tag{7.10}$$

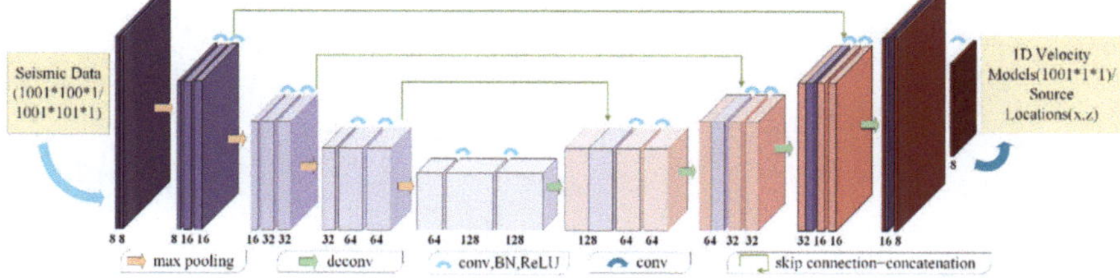


**Fig. 7.22** Network architecture

Here, *Net* refers to a neural network based on FCN, $x$ and $y$ denote the input and output of the network, respectively; $\Theta = (K_1, K_2, b_1, b_2)$ represents the parameters to be learned by the network, where $K$ is the convolutional weight and $b$ is the bias; $R$ denotes the nonlinear activation function; $M$ is the downsampling function (such as commonly used max pooling or average pooling); $S$ refers to the softmax function; and $_2$ denotes the convolution operation.

During the training process of the network, it involves learning to map the input seismic waveform data to the corresponding velocity model, which requires fitting a nonlinear function. Therefore, the learning process of the network can be regarded as solving an optimisation problem:

$$\hat{\vartheta} = \arg min \frac{1}{mN} \sum_{n=1}^{N} L(v_n, Net(d_n; \vartheta)) \tag{7.11}$$

Here, $m$ denotes the total number of pixels in a model, and $L$ represents the metric quantifying the error between the actual velocity model values and the predicted velocity model values. In the method presented in this paper, the $L_2$ norm is used to measure the error.

Only after training with a sufficiently large dataset of samples containing labelled information can the network be considered fully trained. Subsequently, waveform data from other velocity models are used to test the network, and the predicted results are compared with those from the true velocity models.

After constructing the velocity model inversion network, the predicted velocity information, together with the waveform data, is used as input data to train the convolutional neural network for earthquake source localisation. This enables the development of a network capable of localising earthquake sources in different work areas even in the absence of prior velocity information. The two trained networks are then integrated to form a joint inversion and testing network. By inputting test data, predictions for both the velocity model and the earthquake source location can be obtained. The prediction results of the network model are shown in Fig. 7.23, and the error distribution of the localisation results for 100 test earthquake sources using the velocity model is presented in Fig. 7.24.

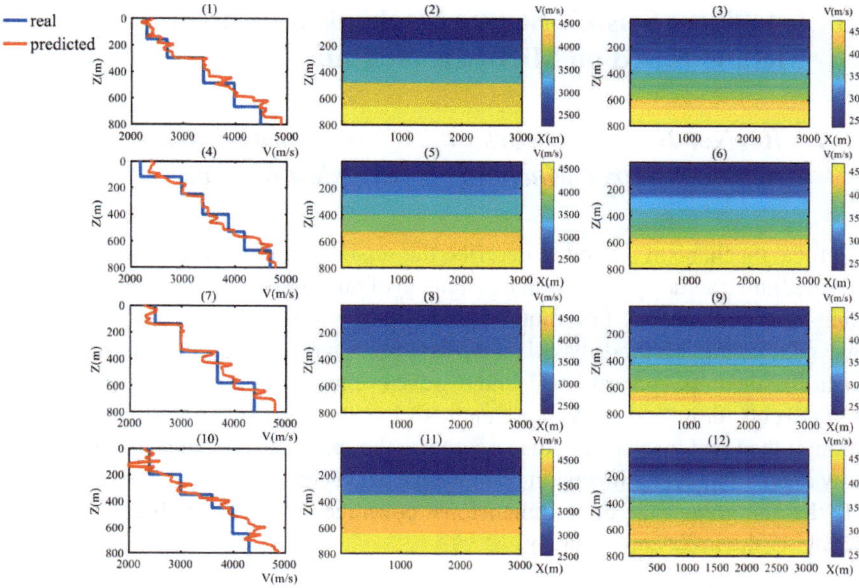


**Fig. 7.23** Prediction results of the velocity model

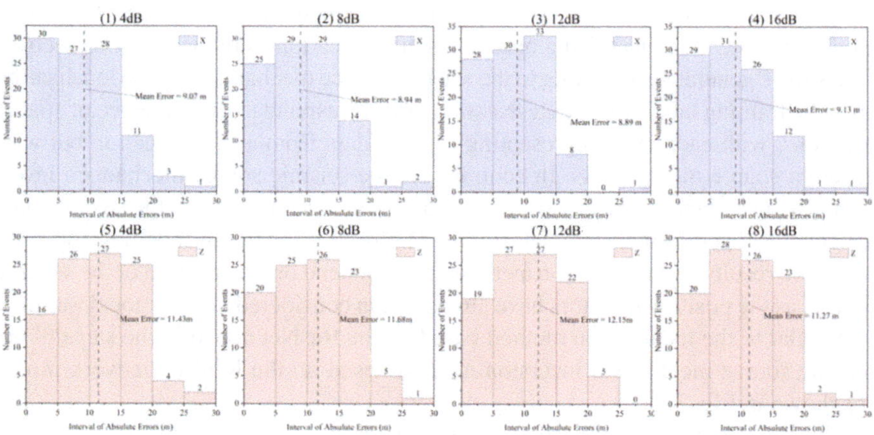


**Fig. 7.24** Error distribution of location results for 100 test seismic sources

## 7.3 Intelligent Seismic Source Mechanism Inversion Method Based on Microseismic Data

### *7.3.1 ResNet-Based Network and Classification-Based Seismic Source Mechanism Inversion Method*

The focal mechanism is used to describe the seismic moment tensor during subsurface medium fracturing. The focal mechanism of microseismic events can be inverted using methods such as first-motion polarity, full waveform matching, and time-reversal imaging. Methods based on the first-motion polarity of microseismic events can utilise the initial polarity of P- and S-wave events, or combine the amplitude ratio of P/S events for inversion. As these methods involve waveform polarity in the calculation of the focal mechanism, discussions on the spatial arrangement of the detector array and its impact on inversion results have also been conducted. The results indicate that the greater the three-dimensional coverage angle of the observation array, the higher the inversion accuracy.

(1) Principle of the Method

Building upon the previous research, the ResNet network has achieved waveform localisation under different seismic source mechanisms. Inspired by this, and considering the reuse of neural network architectures, the method for generating labels was modified: instead of classifying based on source location as in the previous section, labels were generated according to the seismic source mechanism. In the localisation task, the training labels for the ResNet network correspond to the sequence of spatial positions, with each label representing seismic data from a fixed location but with different source mechanisms. In contrast, for the seismic source mechanism inversion task, each label corresponds to a fixed source mechanism, but the seismic data are generated at different spatial positions. The key feature of this labelling approach is that the results of the inversion method do not depend on the location of the seismic source, as the positional factors have already been considered during model training.

Similar to the localisation method based on the ResNet network, the steps of the seismic source mechanism inversion method based on the ResNet network model are shown in Fig. 7.25.

The reverse modelling process, from grid partitioning and the generation of synthetic data to the flattening of P-wave events, follows the same steps as previously described. However, the labelling approach for the seismic source mechanism inversion differs from that used in location methods. For seismic data used in source mechanism inversion, the labels correspond to the source mechanism parameters obtained from the forward modelling. After undergoing a series of processing steps such as down-sampling, windowing, noise addition, and normalisation, the synthetic data are divided into training and test datasets. The training dataset is then input into a neural network, which is trained to map the seismic data to the corresponding source mechanism classification labels. Through these labels, the inversion of the relevant parameters for the seismic source mechanism is achieved.

**Fig. 7.25** Workflow of the microseismic mechanism inversion algorithm

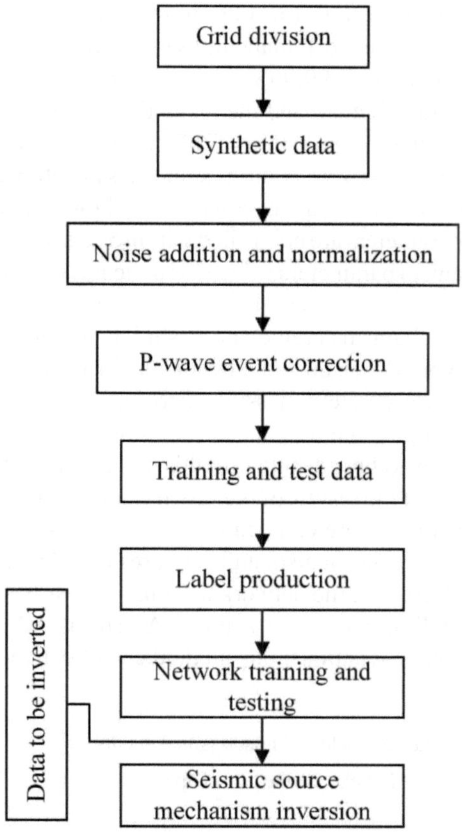


(2) Analysis of Test Results

Inspired by the microseismic event localisation method based on the ResNet network model, this section discusses the source mechanism inversion method based on the ResNet network model. The neural network architecture, training process, and dataset preparation used in both the localisation method and the source mechanism inversion method are identical. The difference lies in the method of label creation: for source mechanism inversion, classification is performed based on the forward modelling parameters (strike, dip, and rake) used during forward modelling. Each of the three parameters is discretised to a certain extent within its respective range, and a sequence of source parameters is formed by combining these parameters in all possible permutations. Each sequence is then assigned a label, and this label is used as the training set label. It is noteworthy that, during the label creation process, the influence of the source location is not considered. Therefore, for the same label, the data correspond to the same source parameters but are not restricted by the source location. After normalisation, down-sampling, cropping, and noise addition, the forward modelling data are divided into training and test sets. The training dataset is used to train the neural network, and the accuracy of the inversion method is evaluated using the test results. The test results show that, under varying degrees of noise interference, the neural network maps the seismic dataset to the correct source mechanism with an accuracy of 99.06%. This demonstrates that the inversion accuracy of the method is satisfactory, and the method also exhibits a certain degree of robustness to noise. Further examination of the misclassified data (Table 7.6) reveals that, although the neural network exhibits some classification errors leading to deviations from the true source mechanism, comparison shows that the errors in the three source mechanism parameters are controlled within 5°, as the discretisation interval for the source mechanism parameters is 5°. This indicates that, under noise interference, the classification errors of the neural network remain within an acceptable range. When the source mechanism parameters are similar, it is almost impossible to distinguish between two types of seismic data by manual inspection. However, a well-trained neural network can accurately classify microseismic data in the vast majority of cases (99.06%), mapping them to the correct source mechanism parameters (Figs. 7.26 and 7.27).

Another advantage of this method is that the location of the seismic source does not affect the results of the seismic source mechanism inversion. The labelling approach adopted here can address this issue, as each label corresponds to only one seismic source mechanism and is not restricted by the spatial position of the seismic source. Therefore, the network training process has already taken into account the influence of different seismic sources. As a result, when dealing with seismic data from various source locations, this method can effectively invert the seismic source mechanism.

**Table 7.6** Selected misclassified results

| No. | Label | Strike (°) | Dip (°) | Rake (°) | Classification result | Strike (°) | Dip (°) | Rake (°) |
|---|---|---|---|---|---|---|---|---|
| 1 | 86 | 45 | 60 | 165 | 61 | 40 | 55 | 165 |
| 2 | 84 | 45 | 50 | 180 | 64 | 40 | 55 | 175 |
| 3 | 74 | 40 | 65 | 180 | 79 | 40 | 70 | 180 |
| 4 | 81 | 45 | 55 | 165 | 82 | 45 | 55 | 170 |
| 5 | 64 | 40 | 55 | 180 | 44 | 35 | 55 | 175 |
| 6 | 82 | 45 | 55 | 170 | 62 | 40 | 55 | 170 |
| 7 | 3 | 25 | 55 | 175 | 23 | 30 | 55 | 175 |
| 8 | 60 | 40 | 55 | 160 | 40 | 35 | 55 | 160 |
| 9 | 64 | 40 | 55 | 180 | 44 | 35 | 55 | 180 |
| 10 | 80 | 45 | 55 | 160 | 60 | 40 | 55 | 160 |

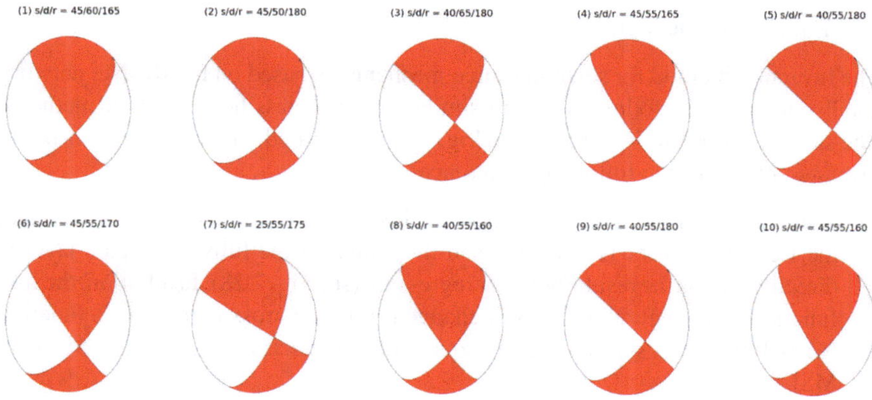


**Fig. 7.26** Beach-ball representation of the true source mechanisms listed in the table

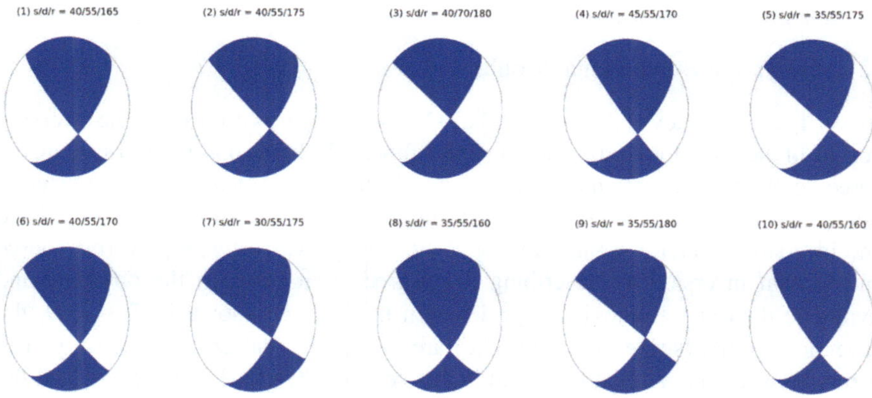


**Fig. 7.27** Source mechanisms predicted by the network for the cases listed in the table

### *7.3.2 Seismic Source Mechanism Inversion Method Based on ResNet and Linear Regression*

In the previous section, the inversion method based on the ResNet network and classification, as well as the seismic source mechanism, achieved favourable inversion results. However, this approach requires prior information to constrain the number of classification outcomes; otherwise, an excessively large number of classes makes it difficult for the neural network to meet practical demands. Furthermore, the classification-based inversion method is essentially equivalent to performing a grid search over the inversion parameter space, with the inversion results depending on the fineness of the grid division. To address this issue, consideration is given to utilising the ResNet network to perform multivariate linear regression in order to directly obtain the inversion results.

(1) Principle of the Method

The inversion method for seismic source mechanisms based on the ResNet network and linear regression follows a process similar to that of classification-based methods. The specific workflow is illustrated in Fig. 7.25. Compared with classification-based inversion methods, the following differences exist:

1) With regard to the neural network architecture, the original Softmax layer in the ResNet network is removed, and the output of the fully connected layer is directly used as the result for linear regression (see Fig. 7.28). In addition, the loss function of the neural network is changed from the cross-entropy loss function (CrossEntropy Loss) to the mean squared error function (Mean Squared Error, MSE).
2) Use the Schmid factor instead of the shear displacement parameters (slip direction, inclination angle, slip angle) to create the corresponding labels.
3) The neural network is trained on all seismic source mechanisms corresponding to a given determined seismic source.

(2) Analysis of Experimental Results

To verify the accuracy of the method, the gCAP software was used to test the inversion results of the same forward modelling data (Table 7.7). In the observation system, the three-component data from nine randomly selected seismometers were subjected to source mechanism inversion, with the seismometer locations shown in Table 7.8 and the inversion results presented in Fig. 7.29. The gCAP software performs source mechanism inversion by describing the degree of fit between the observed and synthetic data and employing a grid search method. As shown in Fig. 7.29, the inversion results (strike, dip, and rake) are 168°, 71°, and 26°, respectively. The errors between the method proposed in this section, the gCAP calculation, and the true values are shown in Table 7.9. The inversion errors based on ResNet are 0.13°, 0.11°, and 0.03°, while those for gCAP are 0.9°, 0.26°, and 4.83°. Both methods yield inversion results close to the true values, but the ResNet-based method is more accurate. The superior performance of the ResNet-based method is attributed to its utilisation of the entire waveform data, whereas the gCAP method uses only part of the data for inversion. Additionally, gCAP only utilises data from nine seismometers, which also affects the inversion results. Since gCAP requires calculation of the

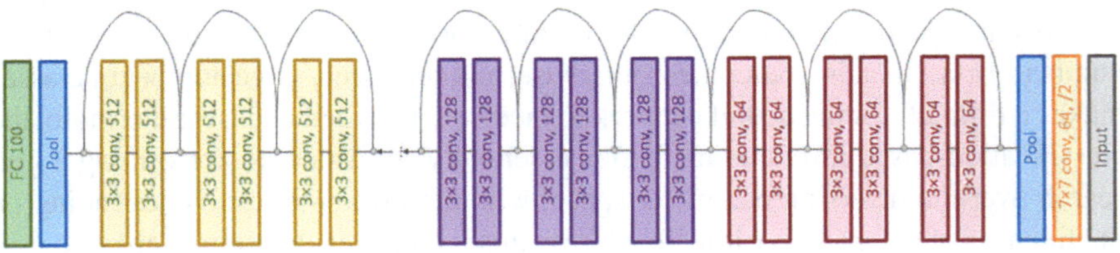


**Fig. 7.28** ResNet network model based on linear regression

initial P-wave arrival time and separate alignment of waveforms along the time axis, processing large volumes of data with the gCAP method involves substantial manual interaction. In summary, the ResNet-based inversion method can conveniently utilise large volumes of seismic data while ensuring the accuracy of the inversion results.

After the network training was completed, testing showed that the prediction accuracy of the network reached 97.65%. Analysis of the simulation experiments indicated that this method achieves high inversion precision, with errors controlled within approximately 1°. To further validate the accuracy of this approach, the gCAP method was employed to perform inversion on the same set of data. Comparative results demonstrated that both methods yield accurate inversion outcomes; however, the method based on ResNet produces more precise results.

**Table 7.7** Fault-plane solutions corresponding to the labels and inversion results

| No. | Label | | | Inversion result | | |
|---|---|---|---|---|---|---|
| | Strike (°) | Dip (°) | Rake (°) | Strike (°) | Dip (°) | Rake (°) |
| 1 | 240.00 | 50.00 | – 90.00 | 239.01 | 50.58 | – 88.82 |
| 2 | 167.10 | 71.26 | 21.17 | 167.23 | 71.15 | 21.20 |
| 3 | 74.96 | 81.35 | – 30.38 | 74.43 | 80.20 | – 30.07 |
| 4 | 190.00 | 40.00 | – 40.00 | 190.26 | 40.34 | – 40.37 |
| 5 | 190.00 | 60.00 | – 10.00 | 190.44 | 60.16 | – 9.69 |
| 6 | 338.49 | 60.50 | – 95.73 | 340.03 | 59.54 | – 95.13 |
| 7 | 280.00 | 10.00 | 130.00 | 281.25 | 10.47 | 131.98 |
| 8 | 280.00 | 20.00 | – 70.00 | 276.67 | 19.72 | – 74.21 |
| 9 | 350.00 | 20.00 | – 150.00 | 349.39 | 20.13 | – 150.07 |
| 10 | 350.00 | 40.00 | 30.00 | 350.37 | 40.09 | 29.91 |

**Table 7.8** Geophone locations corresponding to the data used for source mechanism inversion

| No. | x (km) | y (km) | z (km) | No. | x (km) | y (km) | z (km) |
|---|---|---|---|---|---|---|---|
| 1 | – 1.5 | 0 | 0 | 6 | 1.5 | 1.5 | 0 |
| 2 | 0 | 0 | 0 | 7 | – 1.5 | 3 | 0 |
| 3 | 1.5 | 0 | 0 | 8 | 0 | 3 | 0 |
| 4 | – 1.5 | 1.5 | 0 | 9 | 1.5 | 3 | 0 |
| 5 | 0 | 1.5 | 0 | | | | |

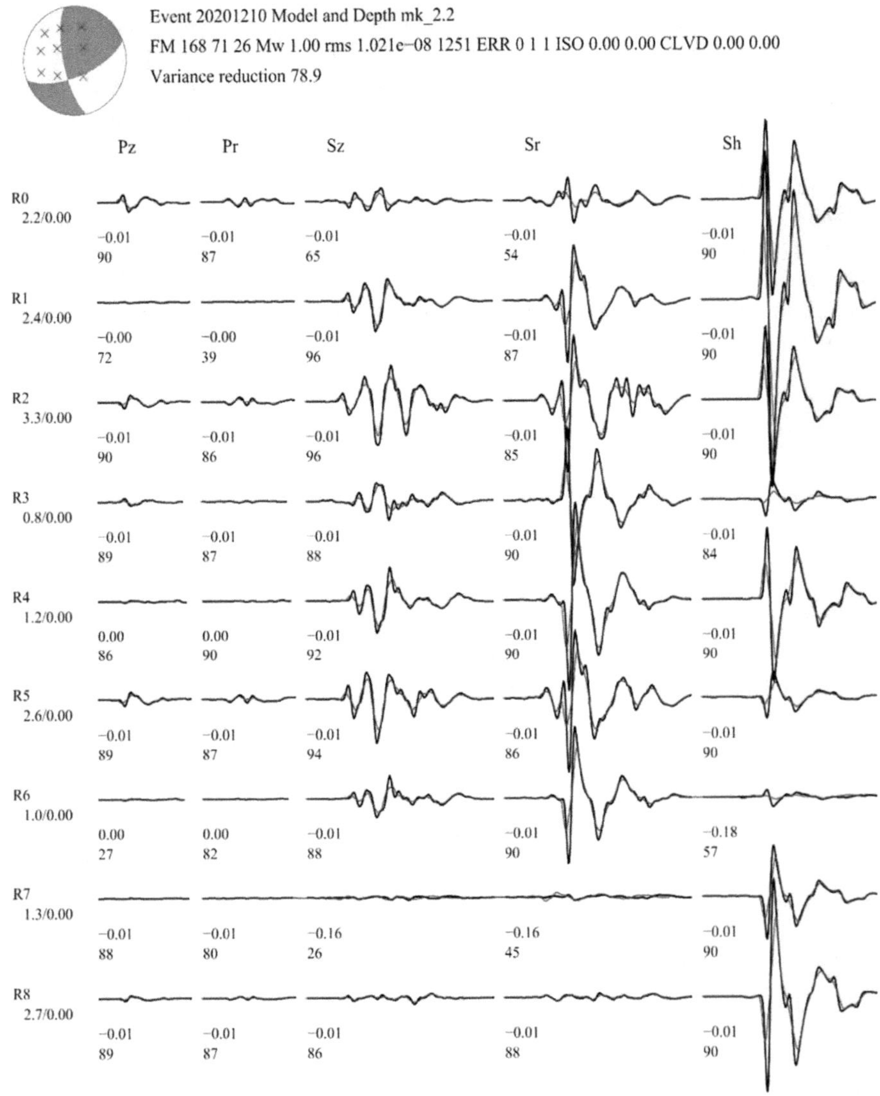


**Fig. 7.29** Inversion results obtained using the gCAP software for the source mechanism corresponding to case No. 2 in the table

**Table 7.9** Comparison of inversion results

| | Strike (°) | Dip (°) | Rake (°) |
|---|---|---|---|
| True value | 167.10 | 71.26 | 21.17 |
| ResNet inversion result | 167.23 | 71.15 | 21.20 |
| gCAP inversion result | 168 | 71 | 26 |
| ResNet inversion error | 0.13 | − 0.11 | 0.03 |
| gCAP inversion error | 0.9 | 0.26 | 4.83 |

## 7.4 Section Summary

In relation to the safety monitoring of $CO_2$ geological storage, this section introduces well-ground integrated microseismic technologies. Starting from microseismic data preprocessing, seismic source localisation, and seismic source mechanism inversion, and incorporating deep learning theory, an efficient and highly accurate intelligent computational method has been developed.

The introduction of deep neural networks has enabled intelligent picking of arrival times and seismic phases for high-frequency single-component acoustic emission signals, three-component signals from surface sparse stations, and single-component elastic wave signals in boreholes. Actual data tests show that the accuracy of arrival time picking can exceed 90%, and the accuracy of P-wave picking can reach as high as 98%. The system systematically investigates the kinematic characteristics of borehole microseismic signals and proposes a directional vector median filtering algorithm for denoising and enhancement. Based on the dynamic characteristics of microseismic signals from surface sparse stations, a denoising algorithm based on residual learning has been developed, which improves the detection probability of weak signals. Centred on the above key technologies, an intelligent microseismic signal processing technology based on intelligent identification and enhancement has been established, providing effective technical support for the safety monitoring of $CO_2$ geological storage.

## References


1. Zheng J, Jiang T, Wu Z et al (2021) Application of residual learning to microseismic random noise attenuation. Acta Geophys 69(4):1151–1161
2. Liu YZ, Zheng J, Wang RJ et al (2024) DMLoc: automatic microseismic event location workflow based on deep learning and waveform migration. Seismol Res Lett 95(5):2997–3007
3. Zheng J, Lu JR, Jiang TQ et al (2017) Microseismic event denoising using adaptive directional vector median filters. Acta Geophys 65(1):47–54
4. Zheng J, Zhu G, Liu M (2015) Denoising of vibrator data based on fractional wavelet transform. Acta Geophys 63(3):776–788
5. Jiang T, Zheng J (2020) Automatic phase picking from microseismic recordings using feature extraction and neural networks. IEEE Access 8:58271–58278
6. Zheng J, Harris JM, Li D et al (2020) SC-PSNET: a deep neural network for automatic P- and S-phase detection and arrival-time picking using single-component recordings. Geophysics 85(4):U87–U98
7. Zheng J, Lu J, Peng SP et al (2018) An automatic microseismic or acoustic emission arrival identification scheme using deep recurrent neural networks. Geophys J Int 212(2):1389–1397
8. Zheng J, Shen S, Jiang T et al (2020) Design and analysis of deep neural networks for automatic phase picking from three-component microseismic recordings. Geophys J Int 220(1):323–334
9. Shen S, Zheng J, Sun Y et al (2022) A continuous data acquisition system for real-time monitoring of three-component surface microseismic events. IEEE Sens J 22(21):20635–20644
10. Li D, Peng S, Yang R et al (2022) Joint monitoring of $CO_2$-ECBM using multiple geophysical methods: a case study of Shizhuang Town, Shanxi Province, China. Int J Greenh Gas Control 121:103795

11. Zheng J, Meng L, Sun Y et al (2023) GPU-accelerated 3D rotated-staggered-grid solutions for the microseismic anisotropic wave equation with moment tensor implementation. Int J Min Sci Technol 33(4):403–410
12. Zheng J, Li T, Xie J et al (2024) Improved numerical solution of the anisotropic poroelastic wave equation in microseismicity: GPU acceleration and moment tensor implementation. Geophys Prospect 72(6):2329–2344
13. Li D, Zheng J, Peng SP et al (2023) Hydraulic fracturing-induced microseismicity controlled by rock brittleness and natural fractures in Tongren, Guizhou, China. Interpretation 11(4):T745–T755
14. Chen Y (2018) Automatic microseismic event picking using unsupervised machine learning. Geophys J Int 212(1):88–102
15. Forghani-Arani F, Willis M, Haines SS et al (2013) An effective noise-suppression technique for surface microseismic data. Geophysics 78(6):KS85–KS95
16. Han J, Baan MVD (2015) Microseismic and seismic denoising using ensemble empirical mode decomposition and adaptive thresholding. Geophysics 80(6):KS69–KS80
17. Lv H (2019) Noise suppression of microseismic data based on a fast singular value decomposition algorithm. J Appl Geophys 170:103831
18. Maxwell SC, Rutledge J, Jones R et al (2010) Petroleum reservoir characterisation using downhole microseismic monitoring. Geophysics 75(5):75A129-75A137
19. Chen Y, Zhang G, Bai M et al (2019) Automatic waveform classification and arrival picking based on convolutional neural networks. Earth Space Sci 6:1244–1261
20. McCormack MD, Zaucha DE, Dushek DW (1993) First-break refraction event picking and seismic data trace editing using neural networks. Geophysics 58:67–78
21. Ross ZE, Yue Y, Meier M-A et al (2019) PhaseLink: a deep learning approach to seismic phase association. J Geophys Res Solid Earth 124:856–869
22. Titos M, Bueno A, Garcia L et al (2018) A deep neural network approach to automatic recognition systems for volcano-seismic events. IEEE J Sel Top Appl Earth Obs Remote Sens 11:1533–1544

MIX
Papier aus verantwortungsvollen Quellen
Paper from responsible sources
FSC® C105338

If you have any concerns about our products,
you can contact us on
**ProductSafety@springernature.com**

In case Publisher is established outside the EU,
the EU authorized representative is:
**Springer Nature Customer Service Center GmbH**
**Europaplatz 3, 69115 Heidelberg, Germany**

Printed by Libri Plureos GmbH
in Hamburg, Germany